THE VIDEO PENCIL

Cable Communications for Church and Community

Gene Jaberg
Louis G. Wargo, Jr.

UNIVERSITY
PRESS OF
AMERICA

LANHAM • NEW YORK • LONDON

Copyright © 1980 by

University Press of America,™ Inc.

4720 Boston Way
Lanham, MD 20706

3 Henrietta Street
London WC2E 8LU England

Printed in the United States of America

ISBN (Perfect): 0-8191-1086-8
ISBN   (Cloth): 0-8191-1085-X
LCN: 80-7951

To Miriam and Ann

TABLE OF CONTENTS

FOREWORD

In this work on cable communication, long aborning, we would acknowledge the contributions of those who have been pioneers in the emergence and development of the technology, many of which are cited in these pages. These vanguard folk, and others who have anticipated both the blight and blessing of cable's advent, are not responsible for positions taken in this book. These are ours and for them we are accountable.

We are grateful to the Fund for Theological Education and the Rockefeller Foundation, for a fellowship enabling research on cable in a parish setting. And we express special appreciation to United Theological Seminary of the Twin Cities, locus of much of our preparatory dialogue. A sabbatical leave from the school provided time for advancement of the work. The late Thomas C. Campbell read the manuscript, as did other faculty colleagues, gave encouragement and made valuable comments. Joanne Perrin and Mary Ann Murray, friends from the seminary community, each read the manuscript and proof-read the typescript, and Mary Ann also prepared the index. We recognize their contribution with our thanks, and also that of Marian Hoeft, who carefully typed the manuscript and showed much patience in the process.

"The medium has changed. For many, the television camera has replaced the pen."

Chuck Anderson, *Video Power*

CHAPTER ONE: INTRODUCTION

A discussion of cable communication, remarkable "delivery system" for the visual symbol that it is, appropriately begins with

A Scenario

The time: the second quarter of the 15th Century.

The place: the "keller" of a house in Mainz, Germany. As the scene opens, Johannes Gensfleisch, commonly known as Gutenberg, is walking slowly around a strange-looking contraption, stage center. He seems to be making some kind of final adjustment on the machine and appears to be very satisfied with himself. Siegfried, a friend, bursts into the room.

Siegfried: Johann, Johann! Oh, thank God, I've caught you in time!

Gutenburg: In time? In time for what?

Siegfried: In time to stop you. Johann, you just can't do it. You can't turn on that contraption of yours!

Gutenburg: Siegfried, what are you talking about? Why, this machine...

Siegfried: That machine could destroy the world!

Gutenburg: Oh, Siegfried, come on.

Siegfried: "Siegfried, come on" nothing! Listen. Since you told me about it, I've been thinking about what that infernal thing could mean, movable type and everything.

Gutenburg: Yes, my friend, it could mean marvelous things! It could spread new ideas, bring people closer together. Just imagine, we could gather up everything that happened in Mainz, in--say--the last week. And we could *print* that with my machine and let all the people know about it. We could call it--let's see--ah, yes, I have it, we could call it a "newspaper."

3

Siegfried: (more and more agitated) Please, Johann, please! Stop dreaming and start thinking. Think of what could happen. Just imagine some "nuts" getting their hands on that thing--instant heresy with mass distribution! Or dirty books, Johann, available to children, maybe--and all because of you. They'll say it's your fault, and they'll be right.

Gutenburg: Come, Siegfried, you're exaggerating.

Siegfried: I am not exaggerating! Why, your machine could even cause unemployment, what with all those monks sitting around and no books to copy. Johann, this could spell the end of the whole monastic movement. Then you'd have the church on your back. Johann, I plead with you, don't turn it on!

Gutenburg: Well, those things *could* happen, I suppose. But, on the other hand...

Siegfried: On the other hand? There is no other hand, Johann. The world doesn't need it. It doesn't want it. It will use it for nothing but evil. Your gadget is demonic, Johann. Please, in the name of civilization, of culture, of education and morality--in the name of God, Johann, don't turn it on!

Curtain

While our scenario may lack historical foundations, it can be assumed that Johannes Gutenburg would have encountered such opposition to his invention. These are the queries evoked by the advent of any new communication medium. In our own time, developments in telecommunication technology have been both welcomed and sharply critiqued. Some critics have simply watched warily from the sidelines.

Technologies always have within them capacities for good and evil in ambiguous mixture. Nuclear fission, for example, carried with it the potential for a whole new energy source. The generating of electricity, the heating of home and factory could be accomplished without having to destroy the earth in the process. Now, since the Three Mile Island incident at a nuclear electricity plant, the promised blessing has become for many a curse.

Gutenberg's press, of course, allowed the printing and distribution of pornography and "heresy." It also made possible the production and distribution of the Bible and furthered the democratization of the West. We who have authored this book believe the means of communication always show such ambiguity. Our thesis, therefore, is that the *utilization* of a medium--as well as its nature and function--are critical. "Software," in a word, is as important as "hardware."

In our recent past, new technologies have come our way so rapidly we had little time to consider in advance their implications and portents. In fact, as Marshall McLuhan suggests, our media environment has been largely "invisible" to us, at least in terms of its *meaning* for society and its *effects* on humankind. And, the proliferation of scientific know-how has been aggravated by our peculiar fascination with gadgetry and our historic short-sightedness. In 75 years, for example, the automobile has progressed from a crude "horseless carriage" novelty to a finely-tooled and sophisticated machine which has radically altered our way of life. Seldom, during that development, did we discuss values. Seldom was effort made to determine the radical societal effect such technology might have. Not until recently were questions raised about environmental impact and the depletion of finite resources. And now, perhaps, the queries come too late.

It is not too late, we believe, to react creatively to other emergent technologies, particularly the development of multi-channel telecommunication enabled by cable linkage. The technology is already on us, that is clear. Cable systems in many areas of the country beam 20, 30 and even 40 channels of multi-lingual information and entertainment to home receivers. Quite obviously, that is only the beginning. Dramatic development of the new medium seems clearly to be a matter of time, particularly in light of corollary advances in communication--satellite transmission, for example. We see linkage communication systems, in fact, developing with increasing momentum. It would be naive to believe we could somehow prevent such development. What is open and to be decided, however, is what we shall do with the cable--or, indeed, what it will do with us. For, in the last analysis, these are the options. The medium offers us opportunities to humanize or depersonalize, to use the linkage system or be used by it.

Some have accused us of being "cable advocates," as if the technology could be foisted on church and community by some sort of press agentry. We cannot believe so potentially useful a medium could, in fact, be deterred in its extension. But, our purpose is not to be its naive heralds. Our interest is how cable TV can fulfill its promise to become "The Television of Abundance."[1]

We expect the new telecommunication systems to alter our lives in ways which could make Gutenburg's invention pale by comparison. And the intention of this book is to help us prepare for such dramatic change. Happily, the nature of cable technology's development gives us the luxury of anticipation, a luxury we have not often had. To seize the opportunity will make church and community better prepared in some ways, at least, for such far-reaching changes as are promised in a "wired nation."

We are well aware of the risks of committing ourselves to print and permanence as we write about this emerging medium. In our rapidly changing world, insights quickly become antiquated. In a matter of a few years, perhaps months, the limitations of our work will be manifest. Economic recession, energy shortages, other emerging technologies--all will influence cable development. On the other hand, the medium and its portents demand a hearing now. As we write, cable-TV serves a fifth of the nation's television audience.[2] Of the 73 million households with one or more sets, some 14.5 million receive their signals by cable. Agency people at Young and Rubicam predict that one household out of three will be "on cable" by 1981. And, many see an immediate and rapid expansion of the technology when 30 percent saturation is achieved. The United States as wired nation seems now a foregone conclusion.

Meanwhile, major urban areas rapidly are writing the franchises authorizing the start-up of new systems. Some of these, notably the QUBE two-way cable experiment in Columbus, Ohio, enable us to see--in new ways--the special opportunities and issues of linkage technology.

In any case, we are ready to risk boldly, knowing that a faint heart could prevent us from an enterprise we feel is momentous. To wait for further data to support and flesh out our observations would make us a Moses refusing to leave Egypt until he had seen the Promised Land.

Predictably, future research on telecommunication should embrace a *world* perspective. While we shall not here attempt to deal with cable-TV in that scope, we eagerly support other humanists who are raising our society's consciousness to issues of Third World communication. We see linkage technologies, especially those enabled by satellite transmission, playing a significant role in the emergence of developing nations and subjugated peoples. And we celebrate those who have committed themselves to enabling such developments.[3]

Our attempt to discern some directions for cable communication may make us seem like a pair from the "gee whiz" school--faddist, meddling and removed from reality. We are reminded of the ancient Buddhist parable in which a fish asks a tortoise friend to climb out on land to see what it is like. Returning, the tortoise reports, "Well, in the first place, there's no water there, just air." And the fish asks, "Air. Why, what do you mean?" The problem becomes how to excite others about a technology, the development of which promises to reveal a strange and unique environment, but one certain to be ours. As Sloan Commissioner Kas Kalba has predicted, "In another quarter century it [the electronic community] may become the predominant environment in which we live."[4]

NOTES

[1] The "television of abundance" notion emerges from the study of the Sloan Commission on Cable Communication, *On the Cable: The Television of Abundance* (New York: McGraw-Hill, 1971).

[2] "Cable TV: The Lure of Diversity," in *Time* (May 7, 1979), p. 82.

[3] A resolution adopted by the Twelfth General Synod of the United Church of Christ, meeting at Indianapolis in June, 1979, endorsed a long-term program "to help Third World nations, through their churches, to protect their ability to use and control sophisticated new electronic technologies such as satellite, lasers and computers." *Keeping You Posted* (April 1, 1979), p. 3. Already the denomination's Office of Communication may have affected decisions of the World Administrative Radio Conference, concluded in Geneva in December, 1979. Latin American delegations, in particular, had been briefed by the church office and went to Geneva ready to protect their interests, especially in the allocation of frequencies for fixed satellites, a technology promising to enhance Third World communication development. Such allocations were put off for consideration by WARC members until a series of conferences in the early and mid-1980s, an action which will allow developing nations more time to prepare their strategies.

[4] *TV As a Social Force*, p. 159.

"We now have in our hands the means to change irrevocably the mind-numbing course of television."

Douglas Davis, *Newsweek*

CHAPTER TWO

CABLE COMMUNICATION: THE FIRST GENERATION

Cable communication, which began simply as a means for re-transmission of over-the-air broadcast signals, has rapidly evolved into a new medium. Originally designed to bring television to the hinterlands, cable communication now is moving to the centers of population. The movement has been partially motivated by economic factors. Cable entrepreneurs have seen the possibility of marketing their "product" where the consumer is--in areas of urban concentration. Successful marketing, however, was contingent on providing the consumer with goods and services not available through already existing means. Why, for example, should one subscribe to cable services which could add little to what was already available over-the-air?

This problem has by no means been resolved, as witnessed in some cable systems whose failure to provide novel services has put them in economic jeopardy. The impetus to provide "something new," while in the first instance economically motivated, has very subtly moved cable communication from the status of a media supplement to the status of medium itself.

The technology, now three decades old, invites superlatives from some who attempt to describe it. Even a modest assessment would suggest that cable will profoundly affect all our lives. Ultimately, it can be expected to shape the destinies of people everywhere. Coupled with the satellite, it makes credible Marshall McLuhan's notion of the *global village*,[1] an intimate society of world-wide scope. It summons us to an electronic "town meeting" and portends a more democratic community. Its potential is perhaps limited only by human imagination and initiative.

Still, the cable phenomenon has not generated a great public stir. Unlike the advent of television in the late 1940's and early '50s--which caused many to rise early in the morning to watch test patterns, and motivated others to endure the physical discomfort of watching snow-clouded images move across miniscule screens-- cable has hardly caused a ripple in the common person's life.

Meanwhile, cable enthusiasts greet the medium as a "revolution," advancing on us in evolutionary style, but dramatic in the ways it may serve and shape us. It may be, some assure us, the delivery system for all kinds of goods, services and opportunities formerly

11

unavailable, even inconceivable to the average consumer. Others exult in the opportunity to "re-invent" video, an occasion ushered in by public use of multi-channel facilities. Some see cable making available a kind of "video-pencil," a communication tool as basic and ubiquitous as the grade-school Eberhard-Faber.[2] Still others, mixing the metaphor in their excitement, refer to cable as a "cavernous communications highway."[3] Some, however, are more cautious in assessing its promise, finding cable an enigma. Indeed, the passionate embrace and total acceptance of the new medium has been far from unanimous. Indulging in negative hyperbole, for example, some have labeled cable "a monster!"

Obviously, it is important to view this electronic linkage system with a realistic and critical eye. That, in fact, is our intention. But, to engage in a serious examination of cable communication--both as technology and technique--is to be staggered by the possibilities of both good and ill.

Cable Features

Consider, for example, these dramatic features of cable. In the first place, it offers virtually *limitless channel capacity*. Many existing systems promise 30 to 40 channels. That number can readily be increased to 100, and someday--employing who knows what emerging system (fiber-optics, satellites, computerized retrievals)-- we shall have thousands, perhaps millions of "program" options.

Secondly, cable introduces *greatly expanded services* at levels both essential and esoteric. Currently, for example, some cable systems are supplying programs providing specialized information for doctors, teachers and other professional groups. Many cable systems have dedicated a channel to stock market reports, providing business and industry with up-to-the-minute quotations throughout the day. Still other systems offer English-language courses for ethnic minorities. The auto industry is already using videotape cassettes to supply service personnel with the latest maintenance and repair procedures. The step is small from individual videotape recorders in various auto agencies to an electronically linked "network." Such services directly interest relatively small groups of people, of course, but the value to the wider community cannot be denied. A doctor, better informed on a specific surgical procedure, becomes a better doctor. But, "narrow-casting" of information to neighborhoods and other focused constituencies would be both technically and economically infeasible were it not for the uniqueness of cable.

A third feature is cable's capacity for *two-way communication*, demonstrated now in feedback signals transmitted from a consumer's console, but one day--in an application like the long-awaited videophone--enabling human connections in one-to-one conversations as well as discussion within a group. This feature promises to transform television from a *passive* to a *participatory* experience, allowing viewers not merely to "talk back" to the tube and to each other, but to register opinions on community issues through an instant electronic plebescite. The nation's first full-blown two-way cable system, Warner Cable Corporation's QUBE, which in late 1977 began offering the service in Columbus, Ohio, makes it possible for viewers to react to programming by pressing five "response buttons" on a hand-held console. Viewers' decisions are transmitted via cable to Warner's downtown headquarters, recorded by computer and the result relayed immediately to the audience.[4] Potential uses of such two-way capability promise to affect virtually all aspects of out daily lives. To Kas Kalba the extension of television's capabilities through cable is certain to be dramatic:

> "The days of television as a spectator activity, during which we passively witness the unfolding of a Sunday afternoon football game, a light comedy serial, a Geritol psychodrama, or a national political convention, will be replaced by a television through which we directly engage in the act of learning, shopping, voting and working."[5]

The index of yet other features and consumer services offered by cable and other telecommunication advances is impressive, to say the least. One day soon we may see facsimile print and newspaper transmission, meter readings, shopping at home, health monitoring, satellite and cable mail delivery, home computerized data transmission and retrieval, specialized "networks" servicing narrower audience interests, and dozens of other "utilities"--all delivered electroncially. It is altogether feasible through the wedding of cable, satellite and computer for the data contained in entire libraries to be instantly available at the push of a button or the turn of a dial. One pundit entertains a vision of "Teeny TV," Dick Tracy's wrist receiver appearing as a 20 dollar "throwaway" mini-television set picking up signals from satellites, and this within the decade.[6]

George Orwell's gloomy forecast for 1984 was "Big Brother," a communication system used to enslave humankind. Douglass Cater's more benign vision is of MOTHER, an acronym for "Multiple Output Telecommunication Home End Resources."[7] One might hope that

MOTHER's characteristics (more channels, greater channel capacity, "narrow-casting," interaction by the viewer and world-wide programming in "real time") might also reveal special aesthetic sensitivity, a less competitive spirit, higher creativity and some responsible nurturing for adult freedom.

Cable and the Parish

It takes little imagination to envision how the advent of new linkage media will alter both the structure and life-style of the church as a whole and the parish in particular. To assume that the church will escape the changes to be wrought by emerging technologies is in the first instance naive and in the second dangerous. The church exists within and for society, making it both a shaping and shaped institution. Failure to recognize both functions as legitimate areas of ecclesiastical concern could have two distinct effects. On the one hand, the church and its people can become victimized by the technology. But, more important, perhaps, the church could miss a significant opportunity to further its ministry.

What could these technologies mean to a congregation in the Cable Age? Picture, if you will, a church meeting convened via cable. In addition to the convenience of not leaving one's own home, consider the energy savings involved when people need not drive to a central meeting place. Multiply such saving of fuel by the number of church meetings held every month in the congregations of a community and muse about how substantial such conservation might be.[8] Furthermore, a cable-meeting could now include persons often excluded because of age or disability. This could at least make possible a level of congregational participation simply not now feasible.

Or, again, consider the time and energy invested in the parish paper, a publication often carrying information better termed "history" than "news." Let us assume the congregation would "gather" by means of cable every Wednesday evening for 15 minutes to share with each other matters of mutual interest. The experience would be immediate, the material current, the time-saving obvious, and the communication probably enhanced. Small groups meeting by cable to plan curriculum for the Church School, to study a community issue or explore a biblical theme, even to share intimacies in a "sensitivity" atmosphere all become possibilities. Such ministries via cable will have obvious effect on church architecture in the future. Doodling on paper may begin to reveal some much altered floor plans for the congregation living in the "wired" community.

Such musing, of course, begins to suggest more disturbing
potentialities. We may well ask if the wired Church will mean
more or less intimate fellowship than is now experienced. The
"communion of saints" may or may not be enhanced by electronic
circuitry. Awareness of the *demonic* latent in any human creation
would seem as important as the cataloging of its constructive
possibilities. Lacking public awareness, for example, cable--
becoming another super industry--could transmute into the
advertising "agency" supreme, providing infinite options of video,
at its best innocuous, at its worst deadening and degrading.

In addition, cable could also become a master-tool for the
despot--the autocrat in government, business or church. The poor
and disenfranchised could find themselves residents of an electronic
prison camp, their lives constricted by the strands of the cable
and their lot far worse than their present sorry condition. Public
service possibilities of the system could be submerged--as has so
often happened in our media past--by powers motivated only to
persuade and to profit. "Community" could be further fragmented
rather than fostered, and consumerism instead of public good be
the dominant theme.

These implicit dangers, real as they are, suggest an urgent
need for both congregations and communities to begin immediately a
process of video education and experiment. Since cable systems
are locally franchised and promise local services, community leaders
can have uncommon and in some ways unprecedented influence on the
kinds of services and facilities that emerge. Even in places where
franchises have been written and cable systems are operational,
public influence can yet be effective in eliciting the service
functions latent in the medium.

Some have registered surprise that cable issues become important to church persons and agencies. Obviously, the attitude of
many has been that "this is none of the church's business!" The
hard questions posed by some church leaders have disturbed many in
media industries who resent the intrusion of ethical issues into
what they obviously consider their private preserve. The ancient
heresy of separating life into spheres "sacred" and "secular" has
re-emerged in the day of cable.

The Sloan Commission on Cable Communications indicated in 1971
that with cable "an impressive new instrument of communication" had
become available. Its report concluded on this note: "It remains
for society to employ that instrument wisely and well."9 As part
of that society, the church--alongside other agencies and

institutions--shares the responsibility of fulfilling the
Commission's mandate. The moral implications of cable communica-
tion--as it affects the lives of people, molding and shaping values
and ideas--is clearly an area not only of legitimate ecclesiastical
concern but of vital theological reflection.

The particular responsibility accorded the church *at the parish
level* deserves re-emphasis. For, as we have averred, cable is
essentially a *local* phenomenon. While systems will surely deliver
the slickly-packaged programs of the television professionals--the
kind to which all of us have become accustomed and to some degree
"conditioned"--what cable does uniquely is offer us the opportunity
to be in touch with our community. At one time the Federal
Communication Commission had moved to insure localism by assigning
the right to franchise cable systems to local municipal governments
and requiring that the franchise provide local access channels for
public use. While such "public access" is no longer required by
the FCC, local origination by public groups remains part of the
medium's bright promise and ought to be claimed by community groups
as their rightful opportunity. And, because of cable's potentiality
for local service, parish ministers and congregations--as in the
development of no other medium--have an opportunity to fashion
communication for humanitarian and ecclesiastical function.

The advent of radio and television caught the church somewhat
by surprise, as it did the rest of society. Now, it would seem, we
may have the unusual opportunity to anticipate and influence the
coming of this "new" technology, the importance of which can scarcely
be exaggerated. The church, one of whose prime responsibilities is
the facilitating of healing human dialogue, is clearly called to
awareness and action. In the prophetic tradition, heeding the
immediate past, it needs to alert our world to the presence of
"tomorrow."

Sages have assured us that those who do not learn from history
are compelled to relive it. But, as Alvin Toffler declares, "if
we do not change the future, we shall be compelled to endure it.
And, that could be worse."[10] His warning is that "we cannot humanize
the future until we draw it into our consciousness and probe it with
all the intelligence and imagination at our command."

With Toffler, we believe the future--particularly in regard
to multi-channel linkage communication is neither fixed nor frozen.
Rather, we foresee a cable and satellite future both fresh and fluid,
bringing exciting if risky days in which the *parish* becomes a locus
of influence and power in some special ways.

Cable is Born

Early in 1972, a new day dawned for cable communication. The FCC opened the 100 "top markets" in the country to cable television. For the first time it became possible for major urban centers to develop this new technology. With this development, people began to view cable in its broader possibilities.

Before that time, cable television had largely been seen as a device to retransmit broadcast signals to video fringe areas. It offered viewers outside urban centers an opportunity to enjoy the benefits of television. That prospect alone was enough to excite small-town entrepreneurs, particularly those in the business of television sales and service.

Perhaps, as is the claim of the international film journal, *Sight and Sound,* cable communication really has its inception in Great Britain. Cable networks were in operation early in the 1920s for the relay of radio broadcasts, and "with the spread of television after the Second World War these systems were inevitably converted to carry the new signals." Again, the journal avers that

> "all the experimental research for developing mass distribution of television signals through cable was carried out by British companies, and the first city-wide cable network for television was built by British Relay in Gloucester in 1951."[11]

It is certainly true that cable systems have flourished for years in England and Wales as well as some areas of continental Europe.

Meanwhile, on this side of the Atlantic, Robert Tarleton looked up at the growing forest of television aerials on the mountain above his town of Lansford, Pennsylvania, and mused that just one large antenna might serve the whole community. Those musings in 1949 gave birth to cable communication in the U.S.A., with systems gradually appearing--not only in Lansford--but everywhere in the country where television reception was difficult, or where people had limited viewing options.

People were eager enough to pay a modest monthly fee, often well below their phone bill, for example, for the vastly improved television reception made possible by cable. Tariffs from a thousand or more households garnered attractive revenues for the system operator as well. The costs to wire a community were substantial, of course, but once the hardware was purchased and the cable laid, the expense of operation was fairly minimal. The potential profitability of Community Antenna Television (CATV), as it came to be called, became apparent to many. Where it was feasible, therefore, where signals could be received and the population of an area promised enough subscribers to warrant it, cable systems sprang up like wildflowers.

This early development of cable took place largely in "rurban" areas of the nation. So it was that in 1973, the American Lutheran Church--with a heavily small-town constituency--could report that 57 percent of its members lived in areas served or soon to be served by cable systems.[12] Denominations with a more dominantly urban membership could not say the same, of course. But, there was development of the technology in those medium-sized cities not eliminated by the FCC's "top 100 markets" prohibition. While existing U.S. cable systems numbered only 70 in 1952 and served but 14,000 subscribers, by 1977 some 3,800 systems provided cable service to nearly twelve million homes.[13] That number has now swelled to 15 million.

While in those first cable days most urban centers were not directly involved, the profits promised by large concentrations of potential subscribers virtually insured that the urban market would not long remain undeveloped. The transition from its rural setting to the urban context would necessitate cable's providing options and services not already available to the consumer. It would hardly seem likely that urban-suburban consumers would pay a subscription fee to receive what they could get over-the-air free of charge. It was at this point in cable development that television moved beyond simply a tool for redistribution of broadcast signals to a medium in and of itself, and to the development of programs and services uniquely "cable." Further, when in 1977 the FCC lifted its restriction of "pay-cable," CATV's growth was given another boost, since pay-cable operations promise truly profitable returns on an investor's dollar.

Still, if a revolution were occurring, the dramatic portents of the emerging new medium were not widely perceived. Like the development of prior technologies, the ultimate "shape" of cable and its meaning for society were questions pondered by a handful of futurists among media educators, businessmen and clergy. Meanwhile, in the council chambers of many communities, cable systems were being franchised by officials often totally unaware of the far-reaching consequences of their action. Franchising a CATV system was treated with no more significance than providing a sewer system or granting a variance to build a garage.

The Church Gives Leadership

It is not immodest to suggest that church persons were among the first to see cable's potentiality. Communication offices of some denominations functioned as "early warning systems" for this quiet earthquake. The church press made efforts to alert its readers and at least one denomination began to suggest it might operate some cable systems.[14] In fact, some media professionals discovered that-- in an effort to bring themselves abreast of the movement--church persons were their most valuable educators.

Perhaps the first to articulate the possible effects of cable television and to share with the public its significance was Ralph Lee Smith, who--in 1970--published an extended essay "The Wired Nation" in a special issue of *The Nation* magazine.[15] Smith foresaw a "national communications highway," when cable would stretch across the country. But, he tempered his enthusiasm for the humanizing social effects he envisioned, suggesting that it could not be assumed all such effects would be good. He concluded, however, that "under any circumstances. . .the cable will be built, and the aim must be, through positive policy and intelligent action, to take advantage of its tremendous potentials for social good." Calling for the establishment of a Presidential commission on cable-TV, Smith felt its mandate should be "the development of a plan for creating a national broad-band communication system in the United States in the 1970s."[16]

With the passing years, more and more people found themselves involved in studying, planning and philosophizing about this extension of a familiar medium. And, the literature on cable began to proliferate. A new wave of interest rolled from the plunge of major cities into the cable stream. When the FCC made urban systems legally possible, both the industrial and public service dimensions of CATV aroused widespread excitement. Since that time municipalities (often large suburbs) have either been pressed to write franchises or are anticipating the need to prepare for that event. As a result, cable citizen committees have appeared everywhere, charged to study the issues and advise local governments involved. In a few cases, regulatory commissions were established by state governments, as in Minnesota and Massachusetts. It has been increasingly clear that cable is ushering in a new era of public communication with a unique set of problems, and local government (often pressed by local citizenry) has--in many cases--set about informing itself and making preparation.

And the chance for video to function more fully in the public interest has been widely understood by a growing number of community agencies. Typically, municipal study commissions have included representatives of schools, churches, libraries, hospitals and the Jaycees--as well as local government. When people become aware of a medium which enables video communication both within their group and the community, they often demonstrate real interest in the shaping of the local franchise.

Still, it must be admitted, such citizen groups have often done their work in frustration. They have not easily found others who share their vision. And, they have frequently discovered the groups they represent have little interest in implementing a study committee's recommendations. Perhaps, in most communities, public excitement about cable has yet to be aroused. One of our objectives is to aid and abet the spreading of the news.

Rarely, it would seem, do we have such an opportunity to influence our future for the better as is given us in CATV. Not often can the common person participate in designing the structure and utility of an emerging phenomenon. Nor often, it would seem, will the church parish find its ministries so influential in the community at large. The church in our age is given a unique opportunity.

How Cable Works

Unlike normal transmission of television, cable-TV systems distribute signals (originally transmitted over the air, by microwave or via satellites) through low-loss coaxial cables. At present these consist of two conductors, an aluminum shield and an inner copper wire. Plastic foam separates the conductors and creates the electronic field through which the signals are transmitted. A far less expensive alternative, the so-called fiber-optic system, is now exciting much interest. Corning Glass has created a tiny glass fiber which would replace costly copper as a conduction medium and do so with extreme efficiency. A light beam, passed from end to end of the glass filament, would be capable of carrying 10,000 telephone conversation channels, for example. A 100-strand bundle, the thickness of a coaxial cable, could give a home receiver one or two hundred thousand options. Of course, such a system--in time--may be replaced with yet another means of transmission, though the principle of "linkage" would seem to survive.

While there are some variations in the design of present cable systems, they look more or less like this:[17]

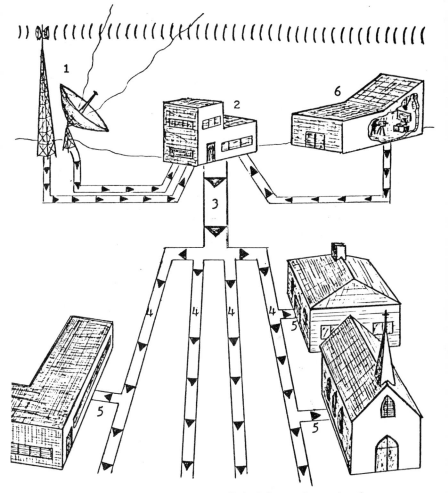

A *master antenna* or *satellite dish* (1) receives signals over-the-air or by means of microwave. This information is transmitted through a cable or microwave to the *head end* (2) an electronic control center where signals are cleaned and amplified and from which films and videotapes can also be originated. This information is channeled through *trunk lines* (3) which distribute signals to a multitude of *feeder lines* (4) branching off the main cable. These, in turn transmit signals to the subscriber's home (office, school, library, hospital or church) through a *drop line* (5) and connection box attached to the regular television set. *Centers for local origination* (6) can transmit live or recorded programs to the head end from other parts of the community and thence to the subscriber's receiver.

For the viewer, installation of the cable means vastly-improved reception. Signals from distant sources become available, and the picture is distortion-free with "ghosts" and "snow" eliminated. Nor need home antennas be installed or maintained. Further, as we have stated above, there is virtually no limit to the number of television signals cable can carry. Even with the technology's infantile development to date, subscribers typically enjoy much greater program selection with cable. And, on that day when cable is linked with a computerized retrieval system, the viewer will have access to a "video bank," providing much of the significant information recorded on film and tape. In Reston, Virginia, a cable system has experimented for many years with the use of home computer terminals, giving subscribers in that community access to hundreds of video options.

The exciting prospect is that viewers can become, more and more, their own "program directors," once the function of a broad- casting professional. Coupled with the consumer use of video cassettes, video-discs and home video recording, cable and satellite linkage systems along with the computer will dramatically increase one's information-entertainment choices, perhaps very soon. Columnist, Steven Brill, is willing to suggest, in fact, that "within the next two or three years, television will probably stop dictating the schedule of even the worst tube addicts."[18] For Toffler, viewers' growing control of their video environment increasingly constitutes a "de-massification of mass media" which "shatters the standardized image of the world propagated [by current communication technologies]. . .and pumps a diversity of images, ideas, symbols and values into society."[19] When video is de-massified, the change is radical--from a medium dominating our lives with few real options, to one controlled by us, involving our selection from a plethora of choices.

Local Origination

Again, unlike over-the-air transmission of television, cable systems require much less expensive transmitting equipment. This means that video information can originate and be sent to the viewer at relatively low cost from multiple sources. So, for the first time, *local origination* ("local O" as it is commonly called) becomes a serious possibility for almost every community in the land. This would include towns too small for feasible broadcasting stations and those sub-sections of larger urban communities which will profit from video slanted to their particular needs. Even in rudimentary installations, cable makes possible people-to-people television, programs of localized interest for local citizenry, produced by their fellow townspeople.

The technology of cable makes local origination possible, until a few years ago regulations of the FCC made provision of so-called "access channels" mandatory. In the top 100 markets, with systems of more than 3500 subscribers, the cable operator was required to provide first three, then one free channel serving local *education* and *government agencies* and for *public access*. Under the public access provision, persons in the community (on a first-come-first-served basis) were given five minutes of free time on camera and were allowed their own determination of the content, without prior censorship. Cable operators argued, however, that the public access channels were little used and the FCC mandate was dropped.[20] Still, the access opportunity can be requested by church and community groups, and--in fact--written as a necessary provision of a local system franchise. Further, particularly where a cable entrepreneur is winning subscribers in a newly-wired community, local programming can be an additional attractive option for the viewer and one more selling point for the systems operator. In addition to opportunities for free public access, of course, community groups and agencies will surely be able to lease channel use, providing other chances for the public to have its video say.

Thus, the way is opened for a truly *public* medium, something quite unique in the American experience of mass communication. On the other hand, the door to that possibility could be slammed shut if the public is not aware of its opportunity and the cable industry side-steps its responsibility.

McLuhan has observed that the *form* of the old medium usually becomes the *content* of the new. Particularly in its early days, television was conceived as "radio with pictures," and old movies still are the content of many hours of telecasting. With cable we can confidently expect many to see the technology in terms of what broadcasting has been and with little appreciation of what "narrow-casting" might be. The popular assumption will be that cable means essentially another transmission form of the television we know about from many years of viewing. Cable system managements, wary of the cost and trouble of providing access facilities and of such intimate dealing with the public, can be expected often to drag their feet on local origination.

That attitude, for example, has been much in evidence at public discussions and cable commission hearings. Industry spokespersons are typically energetic in opposing provisions for local origination. Often discussions have focused on production costs, rehearsal time, professional staff requirements (announcer, director, etc.) and the need for sophisticated production equipment. These are assumed to be necessities in originating programs from local sources.

We are convinced this argument is archaic and naive. Obviously, local origination will have its problems (particularly in our early experiments) and programming worth viewing is going to involve energy and cost. But, over-cautious and inaccurate views of the medium promulgated by some involved in the enterprise could unnecessarily ham-string the promising "new" medium.

Contrast this kind of thinking with that which sees in cable a chance to "re-invent video." Sloan Commissioner Kas Kalba, for example, suggests that television has become "a daily staple of post-industrial living," that it is--in truth--"a basic way of perceiving and communicating."[21] If he is right, and with local origination opportunities enabling more and more people to take hold of video tools, we can expect radical change from that television which has basically been the business of professionals.

Kalba cites numerous examples of video experiment in which exciting change from TV convention is occurring. These can be seen both in what he calls "conceptual video," creating artistic events with television equipment, and "street video," dealing with issues and happenings important to the community. Further, with inexpensive, light-weight *portapak* gear,[22] (shoulder-carried equipment manageable by a single person) and relieved of expected formats, the new video programming "collapses what is conventionally a matrix of professional roles (scripting, directing, sound, lighting, set design, processing, editing, etc.) into as few as one or two individuals."[23] In other words, there is no need for the sizable production crew common to broadcast television.

Part of the genius of cable-TV, we would emphasize, is its *localism*. The authors want desparately to look through the cable-provided video window on our community. We want *programming* (though that may no longer be the best word) of, by and for the people. We would join those who espouse a new "populism" and echo the call for video-power "to the people!" We can do without television's "slickness" in production if what we gain is an open exposure of issues and arts of our community by the very people intimately involved. Though the local high school football team may not take to the field like Pittsburgh or the Vikings, many of us shall watch that game on cable, choosing this video offering ahead of network options because it is *our* town's team and we know some of the kids who are playing.[24]

CATV offers the chance for intimate involvement with video, with the household medium that has been, in ways, irritatingly "distant." It brings on the day of the amateur. It promises *participation* in the creation of television. It does, indeed, put a kind of "video-pencil" in our hands and those of our neighbors.

Clearly, we value participation very much. We take the time, some of our precious leisure, to turn off our television sets in order to golf, bowl, hike, ski, swim, scuba-dive, hunt, fish, camp-- to do many things that *involve us directly*. People will be aroused by a local school-bond issue, they will join the church on the corner, they will seek out the gang at the neighborhood bar, in spite of politics, religion and entertainment dispensed in glittering style by the magic box. Just so, as persons have the video-pencil more and more within reach, we look for them to grasp it with eagerness and employ it creatively.

Intimacy with video--understanding it, using it--will not happen overnight, of course, even when the opportunity is significantly available. Rarely have communities avidly utilized those systems now providing access channels. In fact, some have collected cobwebs from shameful neglect. But, as has been clearly documented, persons educated to cable's origination possibilities can become very excited about the video involvement it offers.

A question sometimes raised about local programs produced for the cable is "Will anyone watch them?" Our belief is that people will certainly watch, that they will be attracted because information of special importance or stimulation of unique relevance is scheduled. Of course, the mass audience of broadcast television will not be available to the video programmers whose purposes can afford to be local and even esoteric. They do not *need* the big audience. Nor are they afflicted with the "Nielson Syndrome"-- that sickness which reduces everything to a common-denominator taste. As a result, we anticipate truly creative things beginning

to happen on the tube via local origination. Graham Wade,
reflecting on a variety of British community experiments, believes that local origination may provide some human portrayals
of unique impact and value. Because the people intimately involved in the issue or experience or art form are producing the
cable presentation, as Wade attests, local-O "has produced
individual pieces of television with a degree of feeling and
insight that broadcast television methods could perhaps never
achieve."[25]

The evidence is that "video literacy" grows as people get
hands-on experience with small-format equipment, in schools, in
professional offices, in churches and homes--even, apparently, in
court rooms. When these video-philes get the opportunity to produce cable programs, we look for fresh illustrations of a lively
new folk art. Not only shall we be attracted to view, we may see
some video that will blow our minds!

James Richards has said that CATV needs to be saved from those
who "think television."[26] His point, an important one, is that
some of the most vibrant cable software will come from those not
trapped by broadcasting's assumptions. Current radio and TV, for
example, have acquainted us with an "announcer," a kind of professional person-in-the-middle. But, in the day of cable, who
needs her or him? Narrowcasting promises more *direct contact* with
the persons and events and ideas that are close at hand. It portends
a town meeting writ large (going on across the land) and writ small
(renewed as a form of local democracy in every cable system). Instead of the mystical possession of a professional coterie, video
symbols may become elements of our common vernacular.

In truth, the voiceless in our society (and many of us are in
that herd) are given both larynx and amplifier in the local origination possibilities of cable-TV. And we (church people, community
groups) dare not be shouted down by those rendered myopic by their
panic over cable economics. Our opportunity may be indeed, the
re-invention of video. And, in the end, the provision of unique
and valued services to cable subscribers may well resolve the fis-
cal dilemmas now troubling pioneers in CATV development.

Two-way Transmission

Yet another important variance with broadcast video is that
cable-TV introduces *dialogue* to mass communication. Cable that
brings the signal *in*, must--by law--be equipped to carry a signal

out. Heretofore, as Richards has recognized,

> "any scale of communication nearing mass has been
> dependent upon one human being's voice. . .being
> heard by large numbers of persons--either because
> of physical abilities to be heard a great distance
> from a stump, by virtue of print or of electonic
> amplification and transmission. [Now, however]
> the technology we have. . .is capable of providing--
> is required by the FCC in the major markets, for
> providing--two-way digital capacity. Two-way audio
> and two-way video technology is available and is in
> use at least experimentally, now."[27]

One of the principal frustrations of the broadcaster and
filmmaker (as for the writer of mass-circulated print) has been
the lack of any immediate *feedback* from the one receiving the
communication. Audience response typically needs to be *inferred,*
on the strength of product purchase, of votes cast for the candi-
date, of box-office profits or the volume and tenor of fan mail.
The eagerness with which feedback is sought, in fact, has given
rise to a thriving industry allied to broadcasting, i.e., the
rating services, which claim to indicate audience size and other
information to networks, stations and sponsors. And, while socio-
logists have consistently challenged the reliability of the ratings--
based, as they are on very small samples of the population--the
surveys of Nielson and the American Research Bureau are read by
broadcasters as "gospel" and programs survive or are junked on the
strength of them.

But, with a two-way system, the picture potentially is radically altered. Feedback could be both immediate and illuminating,
even if such response is at first limited to a switching system,
allowing the use of a yes-no or numbered buttons to indicate pre-
ferences and opinions.

From the viewer's perspective, a two-way system is a true
invitation to involvement. Electric media, in some respects, seem
to have created a hunger for participation. Through their multi-
sensory appeal they have been "involving" and created impatience
with the boring posture of by-stander. But, with all its "coolness,"
its capacity to draw the viewer into the event, broadcast video
has almost exclusively involved us vicariously. Now, our desire
literally to "talk back" to our television set seemingly will be
fulfilled,[28] and the signs of *utilization* of two-way cable are
increasingly evident.

27

More subtle, perhaps, but no less real than industry frustration over the mono-directional system of mass communication has
been the chafing of an audience with no substantial feedback options.
Many have pointed to a kind of *passivity* and sense of *powerlessness*
rife in American society. And, while television may unfairly be
named as chief villain, it certainly has not endowed people with
either power or voice.

Harvey Cox, believing that we are incorrigibly story-telling
animals, and that telling our own individual story uniquely fulfills
us as persons, avers that a "substitute story" is readily supplied
by electric media.

> "The most powerful technologies ever devised churn
> out signals to keep me pliable, immature and weak.
> They hit at my most vulnerable spot. [However] there
> is still time for us to learn again to tell stories--
> mine, yours, ours. If we do not, the signals will
> sweep all before them. Their gentle bleeps and
> reassuring winks will lull us into a trance from which
> there is no awakening. If there is any hell where
> souls are lost forever, that would be it."[29]

Former FCC Commissioner Nicholas Johnson also has railed against
the way our media systems inhibit individual expression. His position is that "individuals need to express themselves--to communicate
to others, to share thoughts and ideas, to build a sense of community,
to overcome the alienation caused by a highly urbanized, industrial-
ized, mechanized life." Johnson looks for a time when the structure
of television "permits and encourages the participation by individuals
with something to say."[30]

That time could well be here. Cable-television will not usher
in the Kingdom. Indeed, it could contribute to further deaden and
degrading of the human spirit. But, the technology *does enable
dialogue,* and--if we labor to assure it--cable *could* evoke human
interchange of a surprisingly vital kind. The availability of
abundant local channels for telecasting and the feedback possibilities of emerging systems do promise new participation in the powerful
world of the media and some political clout to the common citizen.
What advantages cable can bring to groups with humanistic purpose,
what--in particular--it can mean for the church, is a prospect we
can happily entertain.

But, enthusiasm over the two-way possibilities of cable can
lead us to unwarranted optimism. For, significant dialogical

communication, though technically feasible, will hardly happen overnight. Installation of new equipment will be costly, as will the subscriber's monthly rental fees. And well-meaning opponents will alert us to problems which accompany two-way cable.[31]

Economic recession has, in fact, put the brake on this aspect of cable development. For, even the most basic switching systems--potentially providing subscriber yes-no "voting," meter readings and fire and burglar alarm systems--will mean an installation outlay of $50 to $100. More expensive yet will be the sophisticated response systems, including dialing for specific services, retrieving information and ordering pay-tv programs. And, home or office computer terminals (like those operational in Reston, Virginia) will involve expense up to $1,000 for a home system providing banking, shopping, electronic mail delivery and the like. Least likely to be available soon is the video-phone, a device enabling the user both to see and hear the partner in conversation. Estimates are that a city-wide system might cost subscribers from $2,000 to $4,000 for each equipment installation, and that--on a nationwide basis--the cost-per-installation could reach $15,000.[32]

Meanwhile, there are functioning systems currently offering subscribers two-way services. Since December 1977, the much-publicized QUBE system has provided Columbus, Ohio viewers with "participatory television." Warner Cable Corporation, which developed the system, promised QUBE subscribers they would be

"able to give elected officials their opinions, take
college-course quizzes at home, compete from their
living rooms against game-show contestants on their
screens or against other viewers, order merchandize
from stores--all by pushing little buttons on their
home terminals."[33]

Other features enabled by the marriage of computer and television include fire, burglary and other emergency alarm services, as well as 30 channels of information to subscribers who had ordinarily received four local stations.

Toma New Town in Japan offers perhaps an even more revolutionary home TV system that allows viewers to compose their own programs and communicate directly with the local studio. The four million dollar government experiment to realize "upstream capability" links home receivers by computer, coaxial cable and television cameras into what is in effect a "wired city." According to the New York Times Service story,

"viewers can ask the computer for medical advice
through their television sets, or request travel
information on their screen. They can talk to a
studio teacher, or hold images on the screen indefi-
nitely. They can even have their fortunes told on
the screen."[34]

Of course, we can expect technological advance to make such
functions increasingly feasible and less expensive. And, the advantages of such a system are so formidable that, once available,
loud public clamor for two-way communication equipment can be
expected. Clergy and congregations, persons with so much at stake
in a cable technology enabling *dialogue,* can add their voices to
that outcry.

From Conduit to Network

We believe, with Cox, that both the *structure* and *control* of
our present mass media are alterable. Further, we are convinced
that cable makes significant alteration likely, especially should
an alerted church and citizenry seize the *kairotic* moment. Cox
urges an end to further development of broadcast radio and television,
or one-way signals transmitted from a single source to a multitude
of receivers. Instead, he argues, we need

"simple, easy-to-use means whereby small communities,
minority groups, neighborhood unions, and other groups
can communicate effectively with one another. We need
real 'networks,' not the vertical conduits that are
now inaccurately called 'networks.' Ordinary people
need to develop their competence and confidence in the
use of media for their own purposes, to define and
celebrate their own lives and concerns, to deepen their
awareness, to speak out their anger."[35]

It could be that cable is one way to move from the "conduit"
to human "network" style of communication in Cox's vision. The
technology, of course, is no panacea. Yet, we can see latent within
it a means "to help the oppressed. . .claim the promise of liberation,"[36] and to aid us all to realize more fully our true humanity.

30

NOTES

[1] *Understanding Media* (New York: McGraw-Hill Book Co., 1965), p. 34.

[2] We first heard the notion of the "video pencil" from James Richards in 1973 when he was Associate Director of the Office of Communication of the United Church of Christ, in New York.

[3] Catherine Barrett, "The Enigma of a Thousand and One Channels," in *Today's Education: NEA Journal* (Nov., 1972), p. 32.

[4] Harry F. Waters with James C. Jones, "Talking to the Tube," in *Newsweek* (Dec. 5, 1977), p. 107.

[5] "The Electronic Community: A New Environment for Television Viewers and Critics," in Douglass Cater, ed., *Television As a Social Force* (New York: Praeger Publishers, 1975), p. 149.

[6] Richard Rhodes, "80 Ways the Eighties Will Change Your Life," in *Playboy* (Dec., 1979), p. 267.

[7] *Op. cit.*, p. 5. By "real time" Cater means video that is live and un-edited in which the viewer sees events at the time they actually unfold.

[8] Paul Polishuk reviews research and applications on the subject of telecommunications as a substitute for travel in his article "Review of the Impact of Telecommunications Substitutes for Travel" in *IEEE Transactions on Communications* (Oct., 1972), pp. 1089 ff. Ralph Jennings, in an unpublished memorandum from the Office of Communication, United Church of Christ, dated August 2, 1979, sees the energy shortage as "likely to change the ways we are accustomed to handling inter-personal relationships and managing the affairs of our institutions. Modern communications and information systems can help us to keep in touch with one another at a time when travel costs may preclude personal meetings."

[9] *Op. cit.*, p. 172.

[10] *The Futurists* (New York: Random House, 1972), p. 3.

[11]Graham Wade, "Local Cable Television: Alternative or Dead End?" in *Sight and Sound* (Spring, 1977), p. 107.

[12]Robert G. Konzelman, "The Wired Church," in *The Lutheran Standard* (March 20, 1973), pp. 10-11.

[13]Christopher H. Sterling and Timothy R. Haight, *The Mass Media: Aspen Institute Guide to Communication Industry Trends* (New York: Praeger Publishers, 1978), p. 56.

[14]In 1973, the Connecticut Conference of the United Church of Christ, through its Samaritan Communication Corporation, applied for cable-TV franchises in several cities in that state.

[15](May 18, 1970), pp. 582-606.

[16]*Ibid.*, p. 606.

[17]The graphic design was prepared by James Martin, United Theological Seminary senior, from Tomahawk, Wisconsin.

[18]Quoted by Alvin Toffler, "The Third Wave," in *Playboy* (Dec., 1979), p. 180.

[19]*Ibid.*, p. 270.

[20]This represents a tragic repeal of the 1972 FCC regulation requiring separate channels for public, government and educational access. The former regulation was to be fully implemented by 1977, but--according to Louise Sweeney of *The Christian Science Monitor*--"only 100 out of the 750 cable systems in the top-100 market were providing any form of access when the FCC announced the ruling would not have to be complied with." (See the issue of April 19, 1976). Meanwhile, church and consumer groups have petitioned the FCC for new rules that would reinstate the provision of access channels for local groups and provide financing for the community service from pay-cable revenues.

[21]"The Video Implosion: Models for Reinventing Televiison," in *The Electronic Box Office* (New York: Praeger Publishers, 1974), p. 94.

[22]The $2000-$2500 cost of color portapak videotaping equipment puts it within reach of many groups in church and community. Innumerable schools have purchased VTR rigs, which include a shoulder-pack video recorder and hand-held camera--all that is needed to do simple productions for cable transmission.

[23]Kalba, "The Video Implosion," p. 95.

[24]Our informal survey, conducted during an address to a Kiwanis Club, indicated how an important *local* athletic event might outweigh attractive national options if offered to viewers on a cable system.

[25]*Op. cit.*, p. 110.

[26]In an unpublished paper, read at the Speech Communication Association, New York, Nov. 10, 1973, p. 7.

[27]*Ibid.*, p. 1.

[28]For strong encouragement of citizen response to the video industry, see Nicholas Johnson's *How to Talk Back to Your Television Set* (Boston: Little, Brown and Company, 1970). A more recent work informing the public on how to protect its rights in broadcasting is Les Brown's *Keeping Your Eye on Television* (New York: Pilgrim Press, 1979).

[29]*The Seduction of the Spirit* (New York: Simon and Schuster, 1973), p. 307.

[30]"The Coming Victory of the New Television," in *Intellectual Digest* (Nov., 1972), p. 29.

[31]Some have argued, FCC Chairperson Ferris among them, that cable's two-way capability threatens further invasion of our privacy. What upstream capacity can deliver, however, seems to us so significant that we are willing to risk the dangers. The realistic recommendation of the Consumer Federation of America is that two-way systems "be developed in a manner so as not to exploit consumers commercially but rather to meet their informational and social needs. Effective safeguards against invasion of privacy must be considered paramount." The statement is from the Federation's 1978 "Policy Resolutions," p. 6.

[32]Estimates are from *The Citizen's Handbook on Cable Television* (Milwaukee: INPUT, 1973), p. 5.

[33]*TV Guide* (Dec. 24, 1977), p. 3.

[34]*The Minneapolis Tribune* (Oct. 7, 1977), p. 12A.

[35]*Op. cit.*, pp. 313-314.

[36]*Ibid.*, p. 314.

"Life, faith and grace are breaking in today or they are not breaking in at all."

Thomas C. Campbell, *Voices*

CHAPTER THREE

THEOLOGICAL PERSPECTIVES ON CABLE

We disclose our theology by the *form* our communication takes. Our theology shows, as we have said elsewhere,[1] whenever we preach, write, use the arts or mass media. So it will be with cable. The *form* of our cable communication, along with what is overtly stated, will convey a meaning all its own. And, if that be so, we need some theological clarity about what we are doing.

Any human communication act, of course, conveys meaning at several levels. For example, what "doctrine of humanity" is implicit in the television commercial which nudges us to buy a product (or a person, in a political campaign) on the basis of a catchy slogan amidst some video fireworks? A dehumanized "person-as-consumer" is the obvious assumption. For that matter, what view of God is projected by a pulpit six feet above contradiction? The medium is indeed the message, as McLuhan has insisted. Our very style and format *speak*, and a theology comes through *both* in what we say and how we say it.

What, then, have we propagated in our involvement with the media? What theology has been conveyed by church use of radio, television and film? Some clerics have merchandized the Gospel like they were selling soap,[2] as if they were "retailers of optimism."[3] Others have formed "hate clubs on the air" and blasted persons and institutions at variance with their doctrine. Meanwhile, by contrast, main line denominations have been niggardly in their spending of energy and resources on media ministries. Confused listener-viewers are given the notion that God is the "Great Huckster in the sky," quite arbitrary about bestowing favors. Or they may be led to believe mass media are unworthy of the church's attention, that such mundane things bear a taint offensive to the deity. While neither picture expresses the biblical heritage, we may have suggested a God withdrawn from the world or One deceptive and authoritarian!

It appears to us that church persons have often moved casually into the world of media unaware of the exploitive assumptions operative there. Such naivete may result from the fact that explicitly theological questions about involvement in mass communications are too seldom raised. Among those who have written significantly about media ministries, too few have allowed the theological questions to surface. Rarely has anything been written approaching a full-blown theology for mass communication.[4]

It is that enterprise we would encourage as we reflect on theological approaches to cable communication. We are content to suggest directions. We attempt only to single out certain strands from which a more ample theology for cable might be woven. Our intention is to stimulate self-conscious theological reflection in the very activity of cable ministries.

The traditional categories of theology, of course, might well provide the basis for rumination on cable. Our choice, rather, is to let some of the special features of linkage technology suggest directions. To us, then, theological reflection moves from the *dialogical, multi-media* and *process* orientation of the developing medium. Our discussion underlines the reciprocal, two-way nature of the divine-human encounter, it looks at revelation beyond a focus on the "word" and it sees creation as on-going with the Creator as dynamic participant in a changing universe. Our selection is subjective and is meant to be suggestive only. We have been unashamedly eclectic and fragmentary in our discussion, and by no means have we exhausted the issues significant for cable-TV. Still, we believe, these are important considerations in such an exploration. Our hope, too, is that readers will be encouraged to bring their own perspectives, for these will be illuminating to a church shaping a thoughtful reaction to the technology.

The Dialogical Encounter

Communication seen in categories of address and response seems germane indeed to a technology uniquely capable of two-way exchange. And, the reciprocal relation of address-and-response seems to us a dominant--if not exclusive--biblical paradigm of communication, God's with humankind and between person and person. As Hendrik Kraemer phrased it, "The Fundamental vision of the Bible is that God . . . has created man [sic] for communication with him, . . . for dialogue with him"[5] The motif of dialogue, Kraemer notes, pervades the scriptures. It is as though God wants personal relationship, seeing it as the fulfillment both of Divine and human existence. As Daniel Day Williams puts it, "The love [we can say this about communication] between God and his people is given and received on both sides."[6]

So, in the Old Testament, Yahweh seems ill-content with one-way address. Desiring reciprocal communication, God creates Adam and Eve as speaking, reacting persons. They are invited into conversation, as it were, both listening and speaking in their relation with the Creator. That, again, is the posture God assumes--both of speaker and listener--endowing persons with the freedom to choose whatever response they will make.

To the New Testament community, Jesus becomes the very Word of God, and it is a Word of some clarity. For God's self-communication in Jesus of Nazareth seems neither authoritarian or coercive. While Jesus' style is not unambiguous, his encounter with others is often invitational and evocative; in the main he raises issues rather than making bald pronouncements. Cox notes that

"even his so-called sermons are seen by biblical
scholars as later assemblages of single utterances
arising out of particular human situations. They
are not abstract discourses delivered to faceless
crowds without reference to particular events . . .
The style and manner of Jesus' communication closely
coheres with the content of his message."[7]

The same point is made by the Jesuit, Avery Dulles, who reminds us that Jesus'

"use of speech was quite different from the methods
of kerygmatic proclamation. He taught by exemplary
actions, stories, parables and pithy aphorisms. He
preferred to ask questions rather than answer them;
he avoided abstract terminology and rigorous logical
demonstration. By appealing to the imagination and
feelings of his hearers he elicited a high degree of
audience participation."[8]

But, communication in that style, one which makes the "other" a participant and partner in an event, opens one to great risk and the possibility of pain. For, in the freedom of the dialogue, where two persons meet, the outcome is never certain. The "word," one's self-disclosure--given in trust--may be met with rejection, ridicule or outright attack. Prophets and apostles (certainly Jesus, himself) refused to compromise what they were saying by making decrees in the manner of despots and oppressors. But, in consequence, they found themselves open to ridicule and persecution. Authentic encounter with another person always involves risk and danger. Intimacy may make one vulnerable; as Jesus discovered, vulnerable unto death!

To Cox, the crucifixion "eloquently expresses the logic of the biblical view of communication." for,

39

> "the mode of God's self-disclosure to man [sic], in
> the life of a man who is abused, rejected and murdered,
> is not accidental to the content of that disclosure.
> God shows Himself [sic] to be one who is willing to
> risk the most dangerous consequences of dialogue in
> order to make His message known."9

A far more comfortable style of communication is to eschew dialogue and take cover in one-way exchange. But, as Martin Buber averred, to do that is to turn our backs on what makes us distinctly human. In effect, I become truly in intimate relation with *thou*. And, while life forces us again and again to resort to overlapping monologues in the guise of communication, to be content with such ersatz intimacy, to exist wholly in the world of *I-it*, is to be no longer a person. Moreover, to see the theological dimension, Buber insists that God's encounter with us is never apart from such moments of dialogue, that my trusting and genuine relationship with you affords us the "glimpse through to the Eternal Thou."10 Partners in the closest communication, we find significance that transcends the mundane, we find the meaning of existence in both its breadth and depth. We have an experience of human fellowship so fulfilling that it gives us intimations of the Holy.

Human interaction, again, seems essential to genuine communication and is a part of God's creative provision for human freedom. Arguing for a dialogic view of the process, John Bachman labels one-way transmission as "manipulative approaches" that are "unworthy, inappropriate, and ultimately ineffective." His point, stated in theological terms, is clear:

> "The Christian incarnation symbolizes a relationship
> which is much more conducive [than "transmission"]
> to communication. God in Christ has entered into
> human affairs, not imposing His [sic] will on persons
> but respecting individual integrity. He demonstrates
> the necessity of 'participatory relationships' in
> communication."11

But, as we have observed, mass communication is a unidirectional process. It has offered address but has not facilitated response in reciprocal dialogue. The receiver of communication has thus become the consumer, often perceived by the communicator as one to be narcotized and manipulated. In truth, mass media have often been obstacles to communication and have functioned mainly to coerce and manipulate their audiences. Of course, the church, too, has been

guilty of preaching and teaching in monological and coercive
fashion. But, when it does so, the Christian community imperils a
principle at the heart of its tradition--that faith is an act of
free persons, that God never violates our "sacred" power of choice,
the power--in fact--to reject the Divine if we will.

In its ministry to mass media and through them to society,
the church needs again to see itself as *advocatus homini*, the
advocate of humanity and the exemplar of communication as communion.

Demonstrating communication which enables a dialogue that heals
and ennobles, the church will then look for methods and media which
facilitate that process and foster community. It will, for example,
be concerned that cable-communication develop its two-way potentiality to the fullest. It will remind the cable industry of this
feature's humanizing capability and encourage its creative utilization. It will foster the community-building uses of electric
technology, cable and other low-cost, easy-to-use hardware (small-
format video recorders, Super-8 movie cameras and closed-circuit
transmission). For, this gadgetry gives us means by which groups
in our neighborhoods, especially minority groups--all those who
want a share of the power--can begin to articulate their identity
and make significant contact with each other. In addition, the
church will make a strong critique of the "anti-communication" so
often portrayed in the images and values of mass media.

Cox may assign too much guilt to radio and television and reckon
too little with other societal sources of monologue and manipulation.
Nor does he credit the media with the role they have played in opening up dialogue on issues of importance.[12] Nonetheless, we believe
him right in calling for alteration of the structure and control of
media--the kind of change that becomes our opportunity in the rein-
vention of video afforded by the coming of cable technology.

We also believe him appropriately sensitive to the ways in
which--in Western society and especially in the Third World--mass
media have fostered a one-way information flow, largely coercive
in its intent. With him we affirm that

"a theology of mass communications media will expose
the fraudulance of anti-communication, alert the
victims of non-dialogical propaganda, and demonstrate
in its own style a risk-taking, response-including
approach."[13]

That style, manifest in the give-and-take of a dialogical modes of communication, has a special opportunity in the day of cable. For this technology, opening up two-way contact between persons and inviting the non-professional into the role of communicator, this "video-pencil" can write the scenario of involvement and intimacy and fulfillment. Cable-TV can be seen, in theological perspective, in its *kairotic* possibilities. It could be an occasion when the light of the Kingdom breaks through into human experience.

Revelation as "Multi-Media"

A second facet of a theology for cable communication in our judgment, accents the multi-media character of God's self-revelation. Since Gutenberg, the interest of the church has fastened on the word as the revelatory instrument--the Word of God in print and the Word proclaimed from the pulpit. That focus, appropriate for its time in history, needs to be widened in the context of an audio-visual world with quite different experiences and expectations. Citizens of the electric "Global Village" may still entertain faith, but--as one has said--they are probably more accustomed to seeing than believing!

We find that larger understanding of revelation in the biblical story itself. To us there seem many indications within the Hebrew-Christian tradition that the divine disclosure in the world engages the whole human sensorium. For example, the occasion of theophany in the Old Testament is described as a multi-media event. The presence and power of God are experienced and symbolized visually, audibly, tactilely and olfactorily. The biblical saga involves the communicatee in the full range of senses, all of which may be stimulated by the symbols of significance. So, there is the powerful imagery evoked by the bush that burns, by cloud and pillar, in garden and desert. And there are the prophetic visions colorfully conveyed by wheels in the sky, through winged seraphim and gilded thrones--as well as the Voice carried on thunder and the whirlwind. And, scenarios accenting touch and smell and taste are there in abundance, as in the story of Jacob's blessing, stolen from the aged Isaac. The blind patriarch is deceived by fur and savory stew as his second son pretends to be Esau. In the Bible, there seem to be no strictures on the means and manner of God's self-disclosure.

The same understanding of revelation in its sensory amplitude seems clear in the way the New Testament points its varied portraits of Jesus. The church has interpreted that life in terms of the doctrine of the Incarnation, the faith that God comes into our midst in the person of Jesus Christ, in very flesh! Basic in Christian

tradition there is the insistence that divine disclosure is a
sensory event to those aware of it. Jesus is the Word of God, as
John's Gospel eloquently expresses it, but it is clear that more
than aspirated syllables is carried in the symbol of the Logos.
Gabriel Fackre, in his restatement of Christology for this visual
and visionary age, characterizes Logos as "the Vision of God."
Employing an abundance of ocular metaphor, Fackre suggests:

> "the biblical tradition . . . is no stranger to the
> imagery of vision--from the sight of creation beheld
> with satisfaction by Yahweh, through the temple
> vision of Isaiah, to the giving of sight to the blind
> by Jesus, and the Johannine vision of a new heaven and
> a new earth. The time of this imagery may well have come,
> not only in terms of its explicit biblical usage, but
> more so in illuminating the center of Christian faith.
> We take this motif-metaphor as a fundamental clue for
> doctrinal reformulation."[14]

Beyond the biblical support for his accent on visual imagery,
Fackre could well point to some of the early Church fathers, many
of whom explored the "vision of God" theme.[15]

Frederick Herzog, another theologian ruminating on the Logos
concept of the New Testament uses communication to convey the
sensory richness he feels carried by the term. In Fackre's trans-
lation the first words of the Gospel of John become "In the beginning was the Vision, and the Vision was with God, and the Vision
was God." Herzog renders it, "In the beginning was Communication."
These provocative retranslations are surely consistent with the
notion of "Immanuel" and penetrate to its depth--God with us in
all the reality of what it is to be human. For, in person-to-person
contact, we are aware of the other as a kind of multi-media
"happening," communicating with us in ways other than speech. In
any active-reactive event between humans, intentional or accidental,
there are many levels of meaning conveyed through a variety of
sensory stimuli. Students of non-verbal phenomena, in fact, claim
that in person-to-person communication some 65% of the transaction
depends on sensory symbols other than audible speech.

We are suggesting that our theology is impoverished by too
exclusive a focus on speech and print as vehicles for revelation.
The medieval church before Gutenberg and, indeed, the Orthodox
tradition throughout its history, employed the richness and color
of icon and chant, of taper and incense, of movement and drama.[16]

43

The mass in its ancient setting must have been in particular a feast for the eyes! Stained glass and mural and banner brought a riot of color into the worship setting. Church interiors of today are certainly pallid in comparison. European cathedrals, impressive as they are, must have lost lustre in a dramatic way as paint and fresco faded and some Reformers and Puritans laid their heavy hands on objects of art.

Again in our time we experience the power of the visual in the ads and icons of popular culture. Ours has been described as the age of the moving image, for visual symbols typically are in motion and constantly changing. Yet, in the act of worship there is often little to excite the optic nerve. Instead, we view a static scene, are inundated with words and find ourselves trapped in a listener's posture by the pew. In Christian education as well our focus would seem overly verbal. We find it remarkable, for example, that a recent publication on resources and methods in adult education could ignore entirely pedagogy through film and video.

Freud insisted that "optical memory residues" and "visual thinking" are fundamental to human beings. Journalist Walter Lippman phrased it this way: "We think with pictures in our heads." The probity of the aphorism seems obvious as we reflect on how we humans conceptualize. Image does seem to coincide with idea. When we express ourselves, it would be very difficult, if not impossible to avoid pictorial symbols--even in our most abstract philosophizing. There is a priority given the visual in the way we employ our senses, a priority seemingly naturally endowed, given in our "creation."

Nonetheless, the very power of the visual symbol, of the moving image especially, is cause for concern. Fred Allen, the radio comic of the thirties and forties, provided a bleak forecast for humankind in the television age. His caustic comment was that we shall evolve into creatures with eyes big as melons (there being so much for the "eye") and brains the size of a split pea (there being so little to excite the brain)! In truth, our machines can sometimes turn us into monsters. On other occasions our inventions can become our gods.

The symbols by which we live, particular symbols by which "transcendence" is conveyed, seem easily to become "ultimate," evoking our worship rather than that to which they point. The Reformers were especially sensitive to this tendency, as were the Hebrew Patriarchs before them. In England, for example, Cromwell's armies defaced statuary and destroyed brass and decoration, leaving shameful aesthetic scars on churches and cathedrals. And the accent on psalter and sermon in the continental church wrought a kind of uni-sensory liturgical event shorn of elements thought to be idolatrous.

Of course, the "demonic"[17] use of the symbol is not so easily avoided. The Reformed tradition tended to substitute verbal symbols for the visual, but who would deny that a new idolatry emerged? In some churches today the pulpit Bible is carried into the chancel by an elder walking ahead of the minister. Meanwhile, the congregation watches reverently in the opening of a worship period starkly devoid of stimulating other-than-audible symbols. Again, to look to another culture, in spite of the Buddha's injunction against veneration of "objects," the Sutras (sacred writings) are openly worshipped in some Buddhist sects. Paul Tillich's reminder that the symbol itself can claim ultimacy is always important to people at worship. But to believe that such idolatry is escaped by suppressing one or another kind of symbol seems a naive hope.

In truth, we shall often attach too much importance to cable and the sights and sounds it transmits. On the other hand, this technology--as every human medium--is a means by which we may be confronted by the power and presence of the living God. So, we believe, the Christian affirmation of Incarnation reminds us. As Dulles puts it, "The rich variety of modes by which Christ himself communicated suggests that the church, too, as an incarnational reality, may utilize all the possibilities of communication at hand in a given culture.[18]

"At hand" in our day is the exciting possibility of a new electronic linkage system among persons, a system capable of carrying powerful, high-fidelity audio-visual symbols. The church, in its involvement in the ministries of cable communication, participating in perhaps the most significant symbol-making enterprises of our culture, demonstrates its faith in the God who reaches to us through all the media of the corporeal world. The ecclesia communicates in what it does with media as well as what it articulates more overtly by electronic means.

Dulles notes that the Decree on Communications of Vatican II exhorts all leaders of the church to "strain immediately and most energetically to use the instruments of social communication effectively in the many fields of the apostolate." Pointing to a more recent meeting of Latin American bishops in Columbia, he quotes their declaration on mass media:

"The spoken word is the normal vehicle of faith--
fides ex auditu (Rom. 10:17). In our times the
"Word" also becomes image, colors and sounds,
acquiring varied forms from the diverse media of
social communication. The media of social communication, thus understood, are a must for the church
to realize her evangelical mission in our contemporary world."[19]

A Process World-View

A third strand from which we would weave a theology for cable
communication is the so-called process cosmology. In fact, a
theology accenting a process or organismic world-view may well include the prior emphases. That is, in the dialogical nature of the
divine-human relation and the multi-media, incarnational character of
God's self-revelation we find particular illustrations of the process
theme. Our exploration of organismic theology is too succinct and
selective. But because we believe the implications of this stance
so productive in considering cable communication, we are willing to
sin boldly.

The process theologians work out of the kind of world-view put
forth by Alfred North Whitehead. Robert Bryant[20] explains the
Whitehead position in these words:

"This world-view seeks to avoid static conceptions of
space and time as well as the "bifurcation of nature"
into spirit and matter. . . .[It] provides a way of
understanding God which enables us to move away from
static categories ("timeless," "unchanging,"
"impassible") which have often been used in con-
ceiving of God to more adequate ways of characterizing
the Divine. The basic statement of Whitehead applies
here: 'God is not to be treated as an exception to all
metaphysical principles, invoked to save their collapse.
He [sic] is their chief exemplification.'"[21]

An organismic outlook perceives God to be involved creatively
with the universe and other entities within it in a dynamic relationship. The Creator participates fully but not coercively in the
midst of a changing universe and in the affairs of human history.
Again, "God is characterized not only as infinite but finite, as
universal but the most concrete, fully actualized individual, not
only as external but also as temporal."[22]

46

Teilhard de Chardin shares this view of an "involved" Deity, as he embraced a theology employing evolutionary motifs. He was much bothered by the notion of an "extrinsical" God, a deus ex machina "whose existence can only undermine the dignity of the universe and weaken the springs of human endeavor." For Teilhard, such a "pseudo-God" gave the lie to the Christian insistence on the Incarnation.[23] Further, as philosopher Charles Hartshorne protests, the notion that God stands apart from the world, that God has no interaction with humankind of a genuinely reciprocal nature, "destroys all analogy between God and creatures, and . . . contradicts the very meaning of worship and related religious ideas."[24] His faith posits a Deity "interacting with all others, relevant to all contexts."[25]

Our Hebrew forebears, too, expressed a vivid consciousness of the *imminence* of Yahweh. If God is *apart* in wisdom and power, God is certainly the One *present* in the most intimate and concrete expressions of concern for the creation. The intimations of transcendence come in the midst of the "everyday." Events of the Exodus, for example, describe God's liberating work in a reciprocal relationship with the chosen people. In the Sinai pilgrimmage, Yahweh is with them in their decision-fraught adventure even as God leads the people onward. Again, the Old Testament is not loathe to speak of God's reorienting the divine intention in ways more appropriate to specific human needs. Thus, in I Samuel 15:11, it can be suggested that God repents of setting up Saul to be king. Again, in the Jonah parable, we find Yahweh's intention altered as condemnation is lifted from the Ninevites.

Norman Pittenger, one of the helpful interpreters of organismic thought to the Church, supports the notion that it has high consistency with the Hebrew-Christian tradition. "A Christian," he writes,

"may be allowed to say that, if ever there were a philosophy which took seriously the kind of portrayal of God which we find in the biblical record, it is the philosophy of process."

For Pittenger, God does not remain God because of divine distance from humankind but in light of "his [sic] capacity for relating himself afresh to every exigence, every human action, every event in the natural order and in the historical sequence."[25] The biblical portrayal, then, defines divinity not as a static but a living entity. Or, as some anonymous sage put it, "God, to me is a verb, not a noun." A contemporary equivalent tells us "God is

47

where the action is!" In Hartshorne's words, "God, or reality itself, is Process-itself, our God now, more inclusively than he is immutable or eternal Being-itself."[27]

Christological interpretations based on process thought are very important, in our view. For, Christ is seen to reveal a winsome, tender God, a "cosmic Lover" who woos other entities, who only invites into relationship, but does not command the loyalty or force the will of other unique entities. As we have averred, the style of Jesus is dialogical and dynamic. To us then, the personal, participating, processive Deity accented in such a Christology helps to create some "intellectual space" for God-talk in a culture that tolerates such vocabulary with difficulty.[28]

The human tendency toward idolatry may lead us to a theology custom-made for our culture, a culture that includes communication media and styles in currency. On the other hand, the society in which theology is always being re-shaped in enormously influenced by communication--media, methods and theories. To us, then, it is no mere coincidence that process theology comes to the fore in our "Electric Age."

Organismic theology appears particularly germane to our interest in the implications of cable-communication for church response to non-Christian ideologies. According to Bernard Meland,

> "The Christian encounter with the faiths of men [sic]
> is to be guided, not only by what Christian faith,
> through its own distinctive witness, knows of the
> redemptive act of God in its own history, but by what
> all men can know and experience of the creative act
> of God through their own humanity."[29]

Thus is Christian exclusivity challenged and respect encouraged for the other ways in which people speak of their experience of the reality of God? The effect is to affirm a faith distinctively Christian while acknowledging and appreciating values and beliefs arising out of another milieu.

What inferences, then, can be drawn from a process view of Christian faith? What does it mean for communication that God is dynamically involved within a changing universe and in human society? One response is that people are summoned to participate in a similar reality, that the *significant* is touched, in some measure, in that context of unpredictable interaction and change where a "commonness with someone"[30] is possible. In other words, what we describe as communication can itself be defined as process, as a cycle of feed-back and feed-forward. In this configuration (often ambiguous in its human expression) there is the promise of closeness and communion.

Again, the Christ who discloses a "cosmic Lover" Deity, reveals also a divine "method," as it were, non-coercive and winsome, in which other "entities" are recognized, both as autonomous and inter-dependent. The cable communicator who affirms such a view will find discomfort with manipulative methods and will look, instead, for more open and invitational cable-casting styles, expressing convictions non-defensively and eagerly inviting the response of listener-viewers. The format best fulfilling such intentions is dialogical, open, spontaneous and interactive--or, at least, reaching for elements and styles that can be so described.

Cable communication, in process terms, will avoid the didactic and heavy-handed, will reach for parable and pun within relaxed and conversational exchange. The New Testament indicates these often to have been the rhetorical devices of Jesus, devices aimed at bringing the other into active participation and involvement on an equal footing. Further, cable-casters--working out of a process theology--will not perceive truth as some kind of propositional statement which one "delivers" to another. Rather, that which is real will be sought in our meeting with another, *between* person and person.

Obviously, to affirm that non-Christian people also experience the reality of God is to insist on a style of human intercourse lacking the manipulative constraints so characteristic of advocacy and persuasion. In our experience, these have been displayed both in media industries and "show-biz" religion. But, if we really understand that others can add richness and depth to our perception of what is truly significant in life, our approach will certainly contrast with that when convinced we are God's anointed spokes-persons and dispensers of truth. The doctrinaire communicators abound on radio and television as they gather today's "Electronic Church,"[31] and they will flock to cable and satellite from all directions. But, as church persons and secular persuaders, we will demonstrate a theology in the very style we espouse. If the program comes off as ex-cathedra pronouncement, for example, even though its themes are love and liberation the communicators reveal to their listener-viewers an authoritarian, immutable God, One who drops fiats out of the blue and finally perpetuates human alienation.

An Alternative Communication Model

While the canons of classical rhetoric are still the bases of much communication theory, of many college courses, and--indeed--of much practical methodology in mass media, interest grows in alternative philosophies. To put it simply, concern is shifting from *communicator* and *message* to *audience* and *occasion*. Communication is seen more and more in terms of dynamism and process.

The editors of a recent anthology use the battleground analogy to describe the debate in the field, suggesting that the combatants are the opposing forces of monologue and dialogue.

"The 'monological' approach, which defines communication as essentially the transmission and reception of symbolic stimuli (messages or commands), finds its classical formulation in the art and science of rhetoric and its characteristic modern expressions in combative game theory, and the repertoires of mass persuasion. . . .The 'dialogical' approach which regards communication as the path to communion and the ground of self-discovery, found its original champion in Socrates and has as its spokesmen today in such diverse currents of thought as religious existentialism, post-Freudian psychotherapy, and sociological interactionism."[32]

Another way to illustrate what is going on in communication theory is to suggest contrasting models. Classical rhetoric could be said to posit a *monologue-persuasion* model, while a *dialogue-relation* paradigm appears to be coming into vogue. The former exhibits interest in one-way forms of communication and aims at persuasion, the winning of assent. The contemporary emphasis, on the other hand, affirms the method of dialogue to effect relation, both in terms of self-discovery and of communion with others. The contrast, put over simply and in obvious hyperbole, is between the style of a huckster and that of a friend. Interestingly, the church in the first centuries showed real suspicion of the rhetoric alive in the Graeco-Roman world, a rhetoric dominated by sophists and cynics and focused on communication as performance and manipulation.

Of course, the models we suggest are neither mutually exclusive nor fully realizable. Rather, analysis of a particular communication event will place it somewhere on a continuum between

50

poles. What is important, we believe, is the re-appearance of a conceptual model contrasting with one persistent from the time of Aristotle and one that seems dominant in mass communication.

An anecdote of Buber[33] illustrates the disparate paradigms. Suppose, first, that two persons are involved in intensive debate, but without real concern for each other.[34] Each is there defensively, desiring only to articulate their own position. Finally, one of the protagonists gives in, admits defeat and offers assent. Contrast that experience with Buber's recollection of a meeting in which

> "the conversations were marked by that unreserve, whose substance and fruitfulness I have scarcely ever experienced. It had such an effect on all who took part that the fictitious fell away and every word was an actuality."

In the midst of that gathering, Buber and another man found themselves in an argument that seemed irreconcilable. But, related Buber, in the end "he stood up, I stood too, we looked into the heart of one another's eyes. "It is gone,' he said, and before everyone we gave one another the kiss of brotherhood.[35]

Recent scholarship in the speech-communication field demonstrates the amount of interest in communication conceived in dialogical and processive terms.[36] Publications, both in title and content, tend to reflect an interactionist, dynamic view in which something more than the winning of assent is the objective. Advocates from other disciplines reflect similar theories. Marshall McLuhan, for example (the vociferous reactionary response to him notwithstanding), has contributed much to an understanding of communication in our contemporary and shrinking world. Particularly helpful to viewing communication in process terms is McLuhan's notion that media, like persons, have "temperatures," being either "hot" or "cool,"[37] "hot," as an adjective modifying both media and methods, describes communication of "high definition," supplying much data. The result is relatively little "involvement" for the receiver of the message. More communicatee participation is elicited by the "cool" approach, for it provides data less amply, of relatively "low definition" and requires the receivers to supply missing details for themselves.

Because the terms are ambiguous, McLuhan apparently has abandoned the use of "hot" and "cool" as descriptive categories, but his point remains valid. There is, indeed, a significant variance between the didactic and rigid style so charactersitic of the

monologue and the evocative and improvisational situation encountered
in genuinely two-way exchange. The contrast is seen in the difference between lecture and discussion, between sermon and seminar,
between performance for and speaking with an audience. In cable-
casting, the distinction is important. Does the programmer seek
to involve the listener-viewer as interactive partner? Are the
circumstances and environment of listening-viewing taken into
account? Or, is the cable-cast couched in a format and style that
creates spectators? Is it *performance* or *process*, satisfying
basically to the program originator or inviting the listener-
viewer into genuine participation?

In a book co-authored with Barrington Nevitt, McLuhan applies
the principle of participatory communication to administrative
styles. *Take Today: The Executive as Dropout* suggests that
successful management in a fast-changing world means maintaining
flexibility and autonomy. In effect, the "dropout" is willing to
work improvisationally, trusting one's perceptions of what is going
on rather than blindly following the maxims of tradition. The
"dropout" is contrasted with the "diehard" executive, who freezes
on the "controls," unable to anticipate and manage change at
electric speeds. "Dying hard is the worst way to keep in touch,"
admonish the authors. "The new way is dropping out: 'He [sic] who
fights and runs away will live to fight another day."38

The point of McLuhan and Nevitt is illustrated in Edgar Allen
Poe's "Descent into the Maelstrom," in which a crisis evokes contrasting reactions in two brothers. Their fishing boat caught in
the maelstrom, a gigantic ocean whirlpool, one brother remembers the
old saw, "In danger, stay with the boat." But the other, fascinated
by their dangerous situation, observes that floating objects in the
whirlpool are drawn into its destructive depths at varying rates of
descent. In fact, as he is quick to note, the water-casks, lost
overboard, do not move downward at all in their orbit round the
maelstrom. So, trusting what he sees, he throws himself into the
water, catching hold of one of the casks. And, while the brother
in the boat is drawn inexorably to his death, the other rises safely
to the surface as the maelstrom spends its fury. The parable, for
executive styles in general, and for cable "management" in particular
is plain. We ought to trust our *percepts*, not always old *concepts*.

It is in participatory styles of communication, we would
reiterate, that persons discover themselves as well as each other.
It is in the authentic *dialogue* that birth of personhood occurs,
that the truly significant about oneself is encountered. From this
perspective, the manipulative monologues of modern media are a

violation of persons and stultifying to the emergence of the fully human. The one-way messages of mass media seem, indeed, to have weakened the human spirit and to have increased our susceptibility to outside control.

Conclusion

So we would proceed in developing our theological statement for cable technology and in drawing its implications for communication theory. We would hope that others will examine cable-TV issues from their differing perspectives. The matters engaged here obviously require probing and amplification beyond the purposes of this work. For us, however, they suggest directions for assessing the ministries afforded by electronic linkage. As we have suggested, all we do in life, every function in ministry, has its roots in theology, a theology often transparent to the observer though not consciously framed by the communicator. Our feeling is that this theology ought to be one we have considered and are able to articulate. We favor bringing our views and values to the medium, rather than letting it impose on us views and values which may be contradictory to the faith we espouse.

Our approach is admittedly "selective." We have chosen issues we feel particularly generated by the advent of cable, though their application is broadly germane to other communication situations. Cable technology, it would seem, gives us a medium uniquely enabling *multi-media, dialogical* and *processive* communication. That the technology has its denomic latencies is also certain. But, the God who engages us in reciprocal relation, who has chosen to confront us in the ambiguity and amplitude of historical events stimulating the whole human sensorium, and who continues interactive in the changing creation, calls us to reflect the Imago Dei. For, we believe, God has created us to live in freedom, amidst flux and vicissitude. Sensitive to the divine presence within that very process, we may be led to discover who we are, how we are made and how we may touch other lives with healing. With Avery Dulles, we find what he has termed the "secular-dialogic" style

> "admirably attuned to the current communications situation. It is able to operate without authoritative texts, clear categories, abstract definitions, deductive inferences, verbal answers and prefabricated messages. It lends itself to informal and spontaneous dialogue, audience participation and personal involvement, shunning the cerebral abstractness and juridical objectivism of earlier theological models."[39]

These are our theological musings as we pursue responsible ministries in cable communication. Our thoughts are tentative, but we have confidence in their direction.

NOTES

[1]Eugene C. Jaberg, "Your Theology is Showing," *Theological Markings,* United Theological Seminary of the Twin Cities (Winter, 1972), pp. 13ff. Charles Rice makes a similar point in *Interpretation and Imagination,* (Philadelphia: Fortress Press, 1970), pp. 60-61.

[2]Malcolm Boyd, *Crisis in Communication,* (New York: Doubleday, 1957), p. 24.

[3]"Retailing Optimism," *Time* (Feb. 24, 1975), pp. 42-43.

[4]There are some, of course, who have helpfully explored theological issues relevant to mass media. Among them are Everett C. Parker, John Bachman, Kyle Haselden, James Sellers, William Fore and Harvey Cox.

[5]Hendrik Kraemer, *The Communication of the Christian Faith,* (Philadelphia: Westminster, 1956), p. 14.

[6]Daniel Day Williams, *The Spirit and the Forms of Love,* (New York: Harper and Row, 1968), p. 20.

[7]*Op. cit.,* p. 311-312.

[8]Avery Dulles, S.J., "The Church as Multi-Media," *New Catholic World,* (January-February, 1972), p. 45.

[9]*Op. cit.,* pp. 308-309.

[10]Martin Buber, *I and Thou,* trans. Ronald Gregor Smith, (Edinburgh: T. and T. Clark, 1937), p. 28.

[11]"Theology and Communication," in *WACC Journal* (No. 2, 1976), p. 14.

[12]Some observers, Margaret Mead and the French priest, Pierre Babin, for example, have credited mass media with helping to raise world consciousness of societal issues, i.e., hunger, peace, racism, etc.

[13]*Op. cit.*, p. 313.

[14]Gabriel Fackre, in an unpublished manuscript, "Envisioning the Doctrine of Christ," p. 3.

[15]Cf. Kenneth E. Kirk, *The Vision of God* (London: Longmans, Green and Co., 1931), in which the Bampton Lecturer cites the interest of Athanasius, Augustine and Irenaeus in issues related to the vision of God; pp. 181, 305 and 313.

[16]A multi-media worship event vividly etched on our memory occurred in February, 1975, in the Russian city of Zagorsk, home of the Orthodox Patriarch. There, in Holy Trinity Cathedral, worshippers came and went as a women's choir droned endlessly a nasal chant and a priest intoned the liturgy. All around were icons, their brilliance penetrating the dim, candle-lit interior. Worshippers, mainly aging peasant women, brushed past us in their preoccupation with reverential acts. Smells of candles and incense mingled with the body odors of the faithful. At noon, many took out brown paper bags and began to eat their lunch. The "sensuality" of the experience seemed complete when one of the women bit into a large onion as though it were an apple.

[17]We use the term "demonic" here in the sense suggested by Paul Tillich in *Theology of Culture* (New York: Oxford University Press, 1959), p. 60. There he says that symbols "become demonic at the moment in which they become elevated to the unconditional and ultimate character of the Holy itself."

[18]Dulles, *op.cit.*, p. 23.

[19]*Ibid.*, p. 44.

[20]We are in debt to colleague Robert Bryant for his discussion of *Process Theology* in an unpublished manuscript, "Various Theological Interpretations of Norms and Their Influences on Seminary Education," pp. 41, 42.

[21]Alfred North Whitehead, *Process and Reality* (New York: The Macmillan Co., 1929), p. 521.

[22]Bryant, *op. cit.*, p. 42.

[23]Pierre Teilhard de Chardin, *The Future of Man* (London: W. Collins, 1964), p. 267.

[24]Charles Hartshorne, A *Natural Theology for Our Time* (La Salle, Ill.: Open Court Pub. Co., 1967), p. 134.

[25]*Ibid.*, p. 136.

[26]Norman Pittenger, *Process Thought and Christian Faith* (New York: The Macmillan Co., 1968), p. 30.

[27]Charles Hartshorne, "Tillich's Doctrine of God," *The Theology of Paul Tillich*, ed. by Charles W. Kegley and Robert W. Bretall, (New York: Macmillan Co., 1961), p. 194.

[28]Noteworthy, in this regard, is the interest in organismic thought expressed by some in the *scientific community*. Theoretical physicist David Bohm, for example, entertains a process metaphysic to express his scientific world-view. Bohm believes the universe "should not be regarded as made up of things but of a complex hierarchy of smaller and larger flow patterns in which the things are invariant or self-maintaining features of the flow. Quoted by John Platt in "Hierarchical Restructuring," *Radical Software*, (Vol. 2, No. 1), p. 28.

[29]Bernard Meland, *The Realities of Faith*, (New York: Oxford University Press, 1962), p. 357.

[30]Wilbur Schramm writes of communication as "trying to establish a commonness with someone," in "Procedures and Effects of Mass Communication," *Mass Media and Education*, ed. by Nelson B. Henry, (Chicago: University of Chicago Press, 1954), p. 113.

[31]A helpful analysis of the big business-show business phenomenon of religious broadcasting is Charles E. Swann's "The Electronic Church," in A.D. (October, 1979), pp. 17-19.

[32]Floyd W. Matson and Ashley Montague, eds., *The Human Dialogue, Perspectives on Communication*, (New York: The Free Press, 1967), p. vii.

[33]While Buber's thought is distinctive from process philosophy, there are obvious regions of concurrence, as observed by Charles Hartshorne's "Martin Buber's Metaphysics," in *The Philosophy of Martin Buber* (London: Cambridge University Press, 1967), pp. 49-68.

[34]Philosopher Abraham Kaplan calls this phony dialogue a "duologue," his word "for all those human occasions when everybody talks and nobody listens. . . .'Duologue,' he says, 'takes place in schools, churches, cocktail parties, the U. S. Congress and almost everywhere we don't feel free to be wholly human.' In his view, a duologue is little more than a monologue mounted before a glazed and exquisitely indifferent audience." The citation is from *Time*, January 24, 1969, p. 52.

[35]Martin Buber, *Between Man and Man* (New York: The Macmillan Co., 1947), pp. 5, 6.

[36]See such works as Matson and Montague, *op. cit.*; Jon L. Ericson and Robert F. Forsten, *Public Speaking as Dialogue: Readings and Essays*, (Dubuque: Kendall-Hunt Pub. Co., 1970); Dennis R. Smith and Lawrence Kearney, "Organismic Concepts in the Unification of Rhetoric and Communication," *Quarterly Journal of Speech*, February 1973, pp. 30-39; and John Stewart, "Foundations of Dialogic Communication, *Quarterly Journal of Speech* (April 1978), pp. 183-201.

[37]McLuhan, *op. cit.*, pp. 22ff.

[38]Marshall McLuhan and Barrington Nevitt, *Take Today: The Executive as Dropout* (New York: Harcourt, Brace and Jovanovich, Inc., 1972), pp. 5, 6.

[39]Dulles, *op. cit.*, p. 45.

"O, it is excellent to have a giant's strength; but it is tyrannous
to use it like a giant."

William Shakespeare, *Measure for Measure*

CHAPTER FOUR

MINISTRIES IN CABLE

In the average American home, the television set is turned on for some 45 hours a week.[1] Add to that the time given to radio listening, reading mass-circulated papers and magazines and movie-going and the investment in mass-media stimulation yields an arresting total. When cable-TV brings us the reality of a wired nation, we can expect that investment to escalate.

Ministers and other busy professional leaders typically find television viewing and other media exposure marginal activities in life. For most people in our society, however, these are *major* preoccupations. Robert Konzelman points out that the amount of time spent on television alone

"is probably greater than the individual involvement
in any other activity in life apart from work and
sleep. For children and youth, the time spent on
television viewing even surpasses the amount of time
in the classroom. We are told that by the time
American youths graduate from high school, they have
spent approximately twice as many hours in front of
a television set as they have in a school classroom."[2]

McLuhan, cynically suggesting that children know they interrupt their education when they go off to school, assures us that such exposure, over time, has changed the very way humans perceive their environment and react to it. Pierre Babin expresses the notion provocatively in his belief that media have wrought an audo-visual person![3] The *effects* of mass communication are most vividly manifest in young people (children of the TV Age), a fact that is clear enough to parents and educators who live and work closely with today's youth. What is also evident is that the under-30 set is not "going through a phase." They are, as some have said, "another breed of cat!" The generation gap may be both a real and chronic ailment.

Still, for the most part, the ministries of the church--particularly those of the parish--tend to go on as though the Electric Age had never come. In spite of the voracious consumption of media fare and the changes radio, television and film seem to have wrought in society, ministries proceed in ways that largely ignore their presence and power. We may avoid scheduling a Sunday afternoon

61

meeting because of televised professional football, but often our recognition of media influence stops there. And, if parish-based activities of the church discount the importance of mass communication, at the regional and national levels media ministries have often been needlessly crippled by disinterest and inadequate funding. Again, with few exceptions, theological seminaries have shown great reluctance to include media studies in their curricula. One seminary president, rumor has it, told his faculty that since television was only temporarily in vogue, he felt no need to give it serious attention.[4]

Everett C. Parker, David W. Barry and Dallas W. Smythe, in their study of New Haven church response to mass communication in the 50's, concluded that clergy "in general were ready to continue to minister as though these media did not exist."[5] There seems little to alter that assessment 25 years later. In fact, as the publication *Arts in Context* is bold to say,

> "Few ministers have any adequate vision of the possi-
> bilities and opportunities which television affords to
> relate theological truths to the vitalities of our
> culture. It is a fact that most church members in our
> land are tuned in to TV commercials for more hours each
> month than they are to their pastors throughout the
> year. And the total cost of television (largely paying
> for its commercials) exceeds the amount they give to
> support their church. In other words, most ministers
> have a very narrow concern for the way in which their
> people spend their "substance"--their time and money.
> And during the 25 years which television has been a
> significant part of the American scene, the denomina-
> tional bodies charged with responsibility in this area
> have done little to confront their ministers with their
> opportunities and responsibilities in regard to tele-
> vision which were being almost entirely overlooked.
> The results is that there is no appreciable difference
> between the TV viewing habits of those within and those
> outside the church."[6]

We are tempted to agree with Os Guinness' gloomy judgment: "With some form of change inevitable," he says, "it is tragic that much of the West generally and much of the church particularly are the main reactionary forces in the world, attempting to prevent change..."[7] In respect to change effected by mass media, the charge may be especially convincing.

Of course, there are good reasons for conservative reaction to change. For the exceedingly rapid mutations of the Electric Age have brought us all some of the misery of psychic alienation. The speed-up in communication has given us little warning for the chaos and disorder in family and society so much the sign of our times. We look at our recent past with regret and nostalgia. As McLuhan says, "Electronic Man [sic] looks back to the noble and austere values of indivdiual and literate culture with a deep sense of loss."[8] Thus, in the church, we need especially to be reminded of our faith in one who is Lord of creation still, one dynamically involved in its continuance and summoning us to creative life in the midst of process and uncertainty. And, patience and compassion for those suffering from the kind of future shock we cite, is our obvious responsibility.

Further, it would be wrong to paint too darkly the picture of negative church response to mass communication. There are, most certainly, ministers who utilize media creatively within their parish, and the effectiveness of some denominational offices has been dramatic and exemplary. We shall document some examples on later pages. Our hope, however, is for an involved ecclesia, at all levels--seeing media not as ancillary to other functions, but integral to the performance of significant ministry in our time. Cable-TV, especially, may enable us to function more effectively in a multitude of ways--both in the traditional ministries of preaching, teaching and caring, and in some services we shall discover and invent. When cars were available, the circuit-riding parson no longer needed to keep a horse! Just so, we believe, CATV will not so much suggest new responsibilities for parish ministers as provide new facilities to carry on ministries long perceived.

In other words, we see ministerial opportunities in cable-TV more broadly than using linkage technology for cable-casting. While that is an obvious opening for the parish, there are other vital ministries related to the cable. In addition to the *ministry of communication* via cable-TV, we suggest there is a *ministry of critique* to cable operators and to regulatory agencies as well as a *ministry to the consumer* of cable fare. Further, we find it difficult to assign any priority to such ministries. We believe them all to be important functions and lively opportunities for the parish as it lives in the cable milieu. Our suggestion, of course, is that such ministries are not just the pastor's responsibility. All in the congregation need to heed the summons to participate in these *things* of *shapes* to come. For in the wired community, every life and every thing will be radically altered. All who affirm *meaning* in an ambiguous world and seek humanitarian goals have a stake in the way life and media styles emerge.

Ministry to the Consumer

Many people are concerned about the mass consumption of media faire common in our society. They worry--with Fred Allen--about pea-brained, melon-eyed persons evolving as the progeny of coming television generations. They see signs, as we do, of people afflicted by the "pop-corn syndrome," a new social disease, the alarming symptom of which is "uncritical acceptance of entertainment."[9] Those afflicted take on more and more the appearance of sponges. Glassy-eyed and inert they are, posed before the magic box, ready to soak up whatever comes along and reluctant to *think* about much of anything!

Donald M. Kaplan sees the illness as psychopathologic, labeling it "television addiction." To him, the behavior characterizing the symptom is not very exotic.

"In symptomatic television watching, the patient--usually in solitude--finds himself [sic] watching the lite television tube, though no particular programs, for hours on end against his conscious will. His watching overrides his intention to perform even ordinary actions such as answering mail and returning phone calls. It soon overrides his intention to go to sleep at a particular hour. There is no internal motive strong enough to interrupt the behavior. Only external circumstances may interrupt it--a standing appointment, and incoming phone call, etc. Otherwise the behavior terminates in sleep. On closer study, secondary symptoms can be detected. In some patients, neurotice television watching is accompanied by waves of guilt and attacks of anxiety. In others, the consumption of food, cigarettes and alcohol goes up; I suspect a high correlation between this symptom and weight gain. Still others have reported stupor, feelings of disorientation and unreality."[10]

The malaise is no respector of persons. Everyone alive in the Electric Age is in danger of being narcotized and mesmerized by the winsome moving image. And, in every constituency, no matter what the nature of ministry, media consumers are present in great numbers. In other words, no church leader can sidestep this kind of imperative, this ministry to the consumer. The threat is dehumanization on a scale vast and disquieting. The summons to the church--in its role as advocatus homini--would seem clear.

For humans to be *persons*, we would insist, they need *participation* in their environment, the opportunity to act upon it (rather than themselves being acted upon), choosing from the options it offers and enjoying life experiences as both feeling and thinking creatures. So it should be in that part of our world created by mass media, now especially in the emerging new cable world. CATV has been described as "the television of abundance," and that is certainly its promise. Our choices from a plethora of channels will be wide beyond anything comparable in present experience. And, obviously, the choices will range from the shoddy and cheap and obscene to cablecasts that enrich and ennoble. We do not call the church simply to support programming for elite and esoteric tastes. Folk and popular arts promise much delight and fulfillment to the cable viewer-listener. There is finally little "wrong" in light entertainment--so long as persons are aware of their experiences, react as thinking as well as feeling individuals and exercise wider options in their program selections.

Perhaps the best way to help each other prepare for appropriate cable-TV usage is to cut media down to size! We need to dispel the mystique and mystery of electric technology and so be rid of its hypnotic power. At the moment people generally seem over-impressed with mass media, perceiving them as glamorous and inaccessible. But, some lustre may well rub off with the do-it-yourself tool which cable can become. The quicker the diminution of media stature occurs the better, for we may then become more selective in what we view and our evaluation can be more biting. In a word, we have the possibility of being more *in control*, less *controlled* by our media creations.

Let us suggest some *means* by which such a *ministry to the consumer* (a ministry aimed at heightening awareness and awakening the critical faculties) can be sought. First, as more and more churches are discovering, controlled experiences in media utilization can furnish inexpensive resources for dealing with the issues of faith and life. Many persons find it fulfilling to share feelings and reactions after film and video stimulation. Some have found the experience a means to renewal of faith. In our judgment, such an enterprise does indeed help people to view and listen in more evaluative, selective ways.

Slide-sound presentations, multi-media productions and film showings have proven their helpfulness in many congregational contexts, including worship, education and fellowship. Especially in the last decade, there have been a rash of books, articles and study guides designed to help the small group in discussion of the faith issues evoked by current movies.[11] Films, recent enough to have impact, (some now available on video cassettes) can now be inexpensively rented for viewing in church or school. Sometimes groups simply go to the local theater for the current feature. The coffee-and-discussion time that follows can prove lively and enlightening.

A layperson who had taken part in such film-discussions one day greeted us at a meeting and began, with growing excitement, to tell us about a movie he had seen the night before. Suddenly, recognizing his harangue, he cut himself short and exclaimed, "See what you've done! Those film discussions have made me a critic!" So, in fact, they had, and that remark--intended as hyperbole--was an unexpected confirmation of the teaching technique.

We know of congregations who regularly schedule such nights for film viewing and discussion. In some cases church groups have participated in community film festivals. And, in seminaries, there is evidence of interest in cinema, its impact on our society and its use in ministry. One of our students, excited by the medium, built an entire confirmation curriculum around films found in local film libraries, most of which could be rented for very little cost. In the Tillichian method of correlation, he used films to raise the real questions of young people which are addressed by the themes of Christian faith. Evaluating his experiment modestly, the seminarian indicated--however--that attendance at his class was very good and that some children brought friends with them!

Certainly our experience with film can provide some models for us in the day of cable-TV. Even before our community is wired, we can begin to make use of the films on television and other video features. The fact that some publications now include notices of up-coming television programs as well as discussion guides is a noteworthy sign.[12] More and more, significant television features are announced far enough in advance for them to be given promotion in the parish--even for them to be used as planned resources for group study. A careful check of weekly television guides may locate programs of merit in time for mention and group utilization. Our own experiments using television features as the occasion for group theological discussion have proved to us that video stimulation can elicit illuminating and exciting responses indeed. Westerns and who-dunnits as well as documentaries and serious drama can provoke significant sharing, as we have often discovered in the seminary classroom. An Oriental exchange student, for example, found good discussion material in the unlikely series *Hawaii Five-O!* Recently, the light-hearted *Mork and Mindy* provided a student with material for serious reflection.

With cable-TV, video materials can be programmed more conveniently, particularly when the technology provides us with video libraries and retrieval systems. Imagine the flexibility with which the church school could utilize video materials, for example, if a catalogue index located the feature and dialing a code number produced the picture. Ultimately, we should have video materials from every field of knowledge and art available to us in seconds. When instant audio-visual resources can be coupled with live responses and reactions, the prospects for growth and learning are bright, indeed. Meantime, perhaps, our creative utilization of media materials will help people become better consumers, choosing with greater care and more penetratingly evaluating what they hear and see.

A second way to help bring about a new media mind-set is to focus small-group study on the particular issues of mass communication. Sermons, too, can include mention of media matters in an attempt at consciousness-raising. Who, for example, is not concerned about media portrayals of violence? Among parents, who could not be stimulated and helped by the study of the effects of television on children? As preparation for the dramatic advent of cable and the prospect of a wired nation and church, who would not benefit from discussions of the new technology? Communication professionals in the community often can be persuaded to participate in such discussions, as can educators and people from the business community. Resources, too, can be obtained from public libraries. We have found librarians particularly interested in cable and helpful in locating material on the issues, understanding--as many of them do--the importance of linkage technology for library functions of the future.

A third consciousness-raising technique in the ministry to the consumer of cable fare is to seek out illustrative material from mass media programs and to use it in one's preaching, teaching and writing. Too often we have encountered clergy who effect a superior attitude to popular culture, who are particularly nasty in their dismissal of television as "not worth looking at!" Our feeling, rather, is that responsible ministry in our time means carrying a daily awareness of mass media with one, like a wallet.

We can be certain that people in our congregations (or whatever ministry constituency) do have such an immediate consciousness of what is going on in radio, television and film. To make use of what happened Saturday night on *Saturday Night Live*, *Love Boat* or *Fantasy Island* as a sermon illustration on Sunday morning is immediately to touch experiences, shared by many in the congregation and understood by virtually all. When Jesus wanted to make his point plain, he reached into the day-to-day, common stock of people's experiences, using metaphors and references with which he felt his contemporaries could identify and which could be vehicles of the truly significant. By contrast, we clergy persons seem over-interested in illustrations drawn from church and clerical experience, from literary classics or from the Bible--even though the latter's setting is an ancient and agrarian world! What seems to be required is a recontextualizing of the sacred tradition in symbols that say *now* and *you*. Such symbols, we believe, can often be suggested by our listening-viewing experiences.

The point of using media as an illustration mine, of course, is that in this way ministers indicate the importance of mass communication and the cruciality of thinking about what we see and hear. When clergy refer to Walter Cronkite or Carol Burnett, they are saying, in effect, "We are TV viewers, we think the medium is important, and we find it one way life confronts us with issues of faith." Pedagogy by example would seem still the unsurpassed way to stimulate growth in persons.

To cite a fourth means of ministering to the cable consumer, we can *point* to significant media events--to the first-rate film soon to be shown, to the TV documentary scheduled for next week, to the up-coming series which explores an important national issue, or one of light entertainment that can be enjoyed as it stimulates. Again, we ought not come off too highbrow in singling out features we think important. For, indications are that people primarily listen and view for the sake of *entertainment*, not to be educated in any formal sense.[13] Nonetheless, we are taught, in the most unlikely programs, picking up clues (most of them not very subtle) about values, life-styles, and behaviors, and about what is worthy of our "worship."

While we should not ignore the educating that happens in the midst of entertainment formats, neither should we neglect the programming of explicit educational purpose. Public television and radio stations function for our benefit, often with very minimal funding. They will provide better offerings as their resources and public support increase, and in this the churches

can play a part. To alert congregations to what is being provided and what are the needs of such communicators, in the end will
increase our choices of quality features and prepare the way for
intelligent selectivity in the day of cable.

A fifth, and perhaps most fundamental aspect of ministries
directed to the consumer, would be the effort directly to raise
the media understanding of the children who will be our *cable
generation*. While there is much hand-wringing and agonized discussion about our children's absorption with television, not many
positive remedies have emerged. We are encouraged, however, by
recent development of curricula designed to teach *critical viewing skills* to children at various grade levels. Church people have
been among the innovators[14] in this kind of consumer education and,
again, the local congregation has the most immediate opportunity
to foster it--with school boards, among parents and teachers as
well as in its own church school.

Researchers have certainly demonstrated that television viewing has its effects on the outlook and sensitivities of young people,
on their inclination to violent behavior, on family relationships
and a host of other psychic and physical phenomena. The church,
now, has an opportunity--one might say a truly *new* opportunity, in
the advent of linkage technology--to exert its influence on our
educational system for the benefit of our community's children.
As we might expect, the issue of viewing skills classes in the
curriculum is an embattled one where experiments are being conducted,
and these far-thinking educators need our support. A thoughtful
rationale for the effort is supplied by Calvert Schlick, superin-
tendent of instructional services in Mamaroneck, New York:

> "Anything which equips students to deal with the
> barrage of information beamed at them by TV is a
> valid part of the curriculum....We should define
> understanding media as a basic skill, identify com-
> ponent parts of the skill and then discover how to
> teach them"[15]

Where video is used in the classroom, it tends to employ three
basic techniques. Children are involved in *analyzing TV content,*
in very fundamental ways at lower levels, and in styles similar to
the study of plays and novels in junior and senior high schools.
Further, children are being taught how to *do video,* using TV
cameras and videotapes to learn the kind of production techniques
which increase their general understanding of media. TV is also
used as a *springboard*--prompting children to discuss, dance, run

or read, in an expanded *direct experience*.

Neil Postman, who teaches media ecology at New York University, argues that video should be brought into the schoolroom as an "object of inquiry" and used to help students "step outside of and above their information environment." And Ben Logan, editor of *Television Awareness Training*, a viewer's guide, says, "It certainly doesn't seem unreasonable to spend some time and money helping people to use that large TV viewing portion of their lives in a less mindless way."[16]

A final suggestion in nurturing a more mature media mind-set is to involve people in creative production of video programming. Nelson Price, who has observed a variety of people in cable workshop situations, discovers them strongly and positively influenced by their chance to create something to be shown on television. Particularly excited by such opportunities are those who might be expected to be most cowed by media--elderly persons, ghetto blacks, the unemployed.

> "New ability to evaluate information is one clear result of the experience. New self-awareness. A new sense of community. New perceptions. They happened because of new media--cable television and the half-inch video tape recorder. They happened because the myth and mystique of television production had been stripped away. Young people, old people, community people, church people were producing their own programs after only a few hours of training before they began to tape."[17]

In this case, and this *is* the case in cable-TV, consumers becomes *producers,* and are forever *changed* in the way they receive media fare. The ministry to the cable consumer and the ministry of cable *communication* overlap. As Price suggests, perhaps the most important change from television as we know it to cable-TV "is the move from sitting in the easy chair in front of big-production entertainment television screens to being the producer behind the camera of community television."[18] That kind of shift, we are convinced, will also mean a dramatic change in the ways people respond as consumers.

The Ministry of Critique

Mason Williams, the gentle television cynic and entertainer-

musician, has said that he will not only *watch* television, but that
he intends "to keep his eye on it!" Those who have a *ministry of
critique* (to cable operators and regulatory agencies) could well
say the same. All Christians have a stake and a share in this responsibility, of course. And, those who lead the church profession-
ally have the role particularly, as a kind of liaison between the
world of cable and the community of faith. If, as we have said,
clergy can help point people to significant events in media, they
may also be able to signal to media professionals the concerns and
critique of the church.

Sometimes in the past (for this ministry has often been taken
up eagerly), the clerical reaction to media fare has been narrow
and pietistic, but the ministry of critique is not fulfilled, to
our way of thinking, by easy moralizing nor by destructive carping!
Often, when the church has responded to the mass communication industry, it has spoken in prudish, superficial opposition to sexu-
ality and violence on video and cinema screen. Other times there
have been objections raised to uncomplimentary portrayals of the
clergy or the request for more refernces to God. Somehow, as Kyle
Haselden suggests, we have often objected to the wrong things, to
petty issues, meanwhile missing the truly "obscene" in our
society--religious, racial and sexist bigotry and "the ingenious
degeneracy of modern war."[19]

Our critique of cable management (and media leadership, more
generally) can find an appropriate basis in the fundamental dictum
governing broadcasting. The law (as enunciated in both of the early
documents establishing broadcasting)[20] declares station owners to
be *licensees,* given the right to broadcast only so long as they
function "in the public interest, convenience and necessity." So
also it should be in the cable industry. Those awarded cable franchises need to know their license to operate a cable system is subject to review and renewal by the community they serve. In ways
the broadcastors have not, they need to give priority to cable's
public service functions.

Radio and TV managements most often have sought to discharge
their public obligation by setting aside program time (usually the
least commercially valuable) for use by educational, religious and
service agencies in the community, and by so-called "public service announcements" in behalf of the Red Cross, traffic safety and
the United Fund, for example. Too rarely has the full impact of
the Communication Act's philosophy been perceived--that media
operators hold their licenses as a public trust and that *all they
do,* every single program element ought to be justifiable as a

public service. Compelling the broadcasters to support their total programming effort with a thoughtful rationale in the end would benefit both media and the consumer.

Sometimes, to be sure, leaders in mass communication have really understood and embraced the public service notion. Such persons should be supported and encouraged, as should all worthy efforts in cable and broadcasting. More often than it has in the past, the ministry of critique should mean noticing media efforts of value and commending the industry. A credible critique will also take into account the many programs and series deserving our plaudits.

On the other hand, we ought not be wary of well-placed criticism of programs and policies which are not in the public interest. Obviously, such evaluation is difficult and persons will honestly differ in the standards they bring to judge the media. John Bachman, in a 1960 discussion of church and media, suggested a *Christian possibility for judgment*. Accepting broadcasting's focus on the audience, he argued, we need to raise this question in relation to every program element: is the listener-viewer treated as *human* or *less-than-human?* Bachman asks if the audience is regarded as human or mice to be routed through a maze? Are they people, or machines to be operated, vegetables to be cultivated? Are they capable of growing and learning, or simply creatures able to react? If listener-viewers are regarded as human they should be aroused and stimulated rather than stupified; their horizons should be enlarged rather than shrunk; they should be exalted rather than degraded.[21]

In a work that appeared a decade later, Kyle Haselden's suggestion was similar. Discussing morality in mass communication, Haseldon judged immoral in media all that is destructive of persons and preventing their growth in freedom. On the other hand, the media fulfill their ethical responsibility when, according to Haseldon, they encourage and enable the transformation of *people* into *persons.*[22]

The criterion becomes clearer as we apply it to some of the common program elements of media--elements sure to be present in cable-TV as well. Advertising on-the-air, for example, can be judged in terms of how it regards the persons who view it. Does it give reliable information about the product, or does it try to influence people to act without thinking? Are we who listen and view motivated to make conscious, intelligent choices or does the ad nurture human weaknesses and encourage us to indulge them? The *person*, acting in freedom, is a choosing, discriminating creature, but the advertiser may be tempted to by-pass our critical judgment to make the sale.

News and entertainment programs can be subject to similar scrutiny. We may be given responsible, reasonably objective reporting in some depth, or news may be superficial, inaccurate, overly brief and biased. Features we watch for sheer relaxation and enjoyment may lack all artistry as they saturate us with escapist happy-endings. But the program intended for entertainment can also stimulate the imagination and deal honestly with human situations, as--for example--the great novelists have done. Again, media may (and often do) expose the listener-viewer to shocking violence. While opinions differ, some would see television and film as major contributors to the violent mood of a society that seems to live by the gun. If mass communicators deserve recognition for some features of excellence, they are to be chided for so much programming that is--at best--a "vast wasteland" for the consumer, and--at worst--truly degrading of the human spirit. Our point is that cable operators (as, in fact, all media programmers) need to be confronted when commercial expediency overshadows public service obligation.

When we critique program "content," of course, we venture into the danger-zone of control and censorship. Cable technology brings this issue into especially sharp relief. There is both bane and blessing to be recognized in the narrow-casting feature of CATV, its capability of offering program choices for very individual tastes and interests. The wider range of options allows consumers to "program" their own viewing, as it were, to choose what they do with the medium. On the other hand, that fantastic array of options seem certain to include everything from the arts in depth to *Deep Throat*. Experiences with public access channels (where citizens have claimed five minutes to "do their thing" without censorship) indicates that the kooky, the bizarre and the obscene will get abundant play on the cable.

Confronted by the depraved or the heretical, the church has always been tempted to call in the censor. When that has happened in the history of Christendom, we believe the church compromised an important principle, violating human freedom in order to defend dogma or "protect" its people. With Haselden we would argue that censorship and control tend to be more harmful to society than beneficial. "When wrongly employed," he writes, "these restraints on man's [sic] right to know all that he is capable of knowing threaten his freedom and consequently his possibility of becoming a moral creature.[23]

Since we believe censorship to be generally hostile to authentic morality, we would warn those communities tempted to constrict the offerings of cable systems. It appears now that basic regulation of cable-TV will be in the hands of local government. And, in general, the more local the governmental agency, the more interest there has been in censorship of movies, textbooks, publications and electronic media. We anticipate city councils to be strongly pressured to write limiting strictures into franchises and that--from time to time--there will be a public hue and cry over "objectionable material" available in the cable.

When the issue emerges, churches in a community can certainly help in clarifying matters and taking a stand. Urging that censorship and control be avoided as much as possible, church people can show how the alternative poses real dangers to that kind of free decision-making that is the basis of true morality and--indeed--of democratic society. Further, they can be active promoters of *good* programming--the artful, the informative, the humanitarian--all those cable features which probe, prod the imagination and excite personal growth.

Cable technology could enable the end of some subtle controls hampering mass media as they now exist. Many communication analysts have noted an inevitably conservative bias in the media, tending-- as they obviously do--to represent what a society is and to conserve its values. A working party of *The National Union of British Journalists,* for example, giving evidence to the Annan Committee on the Future of Broadcasting, noted a strong tendency towards broad establishment attitudes in media. "Editors and broadcasters are left alone," they alleged, "only as long as they do not stray too far from the broad assumptions on which our society is based."[24]

It is possible, on the other hand, that CATV could somewhat restore the "free market of ideas" envisioned by John Stuart Mill, John Milton and other champions of a free-press, libertarian philosophy. For cable could move to equalize the power, giving common citizens greater opportunity to *choose* and to *communicate*. Both opportunities are restricted in a media system which is big business and highly professionalized. There are grounds, we believe, to expect new and exciting ideas to emerge from such a populist medium, and their circulation to be fostered by it. A free society (and certainly the church) has nothing to fear from truly free expression--even though it may at times offend sensibilities. For, the human community is nurtured on the *new* as well as the traditional, and there can be little growth toward justice and the humanistic society except we are open to novelty.

Obviously, those who manage "public" media, whose fare is brought into the living room, have special responsibility for *what* they program *when*. On the other hand, in light of the promise of instant retrieval from video "banks," home playback of video recordings and the like, the cable operator will finally have very little control over what is finally selected for listening and viewing. The day of cable puts the onus on consumers--who are given an enormous region of choice and thus new responsibility for themselves and their families.

Still, the critique to the cable industry can be important and influential. Further, evaluation of a linkage system's service can provide a welcomed barometer of community reaction to the cable operator, particularly if the evaluation represents a cross-section of opinion. We suggest that every cable community should have a Citizen Advisory Council, gathered from a variety of civic groups, including churches. Such a council, a liaison group between the city government and the cable operator and responsible to advise both, needs to be taken with great seriousness--by its constituency and the community.

Again, thoughtful, critical responses from a particular group (within the congregation, for example) can carry weight, as can individual letters and phone calls. When such statements are articulate and moderate, they can indeed by influential.[25] Review articles in the local press (ultimately on cable programs and services) accenting the good and prophetically pointing out the shoddy can be important in the ministry of critique to the cable industry. Cable operators (as all media management) are dependent on public acceptance, so they are likely to be responsive to a community demand for better programming, particularly if some good program models also exist.

Another expression of ministry includes counsel to the regulatory agencies responsible for cable communication--the Federal Communication Commission, in some instances state cable commissions, and--of particularly of concern to parish and community--municipal governments. Since the latter grant the franchises authorizing cable systems, considerable regulatory power is vested in local agencies. The citizen's council we have recommended can serve a much-needed advisory role to local officials, beginning with a study of community needs in relation to cable-TV, should the system not yet be installed. In some cases, such a committee can be organized even before the franchise is written and thus can be instrumental in shaping this vital document.

Unfortunately, many franchises were proffered before there was any cable awareness, so they were written with minimal caution and little insight into the important community functions of such systems. Where enabling ordinances are not yet final, we suggest such matters as these merit concern: 1) *the franchise length*--we believe it should be limited to ten years (then reviewed carefully before renewal), a period long enough for commercial feasibility, brief enough to insure responsible community service from the cable operator; 2) *the assurance of public service functions*--we suggest that these include the allocation of adequate channels for public access, provision of free hook-up, wiring and subscriptions to community service agencies, the possibility of central studio facilities--and an electronic system capable of modification for technological innovation; 3)*the establishment of fair subscriptions*-- we affirm that service rates should put cable service within the reach of the ordinary citizen,[26] with rate adjustment an issue for review by municipal government.[27]

In short, the franchise should aim for a viable, service-oriented system capable of endurance. Everett Parker expresses it this way:

> "Public-spirited citizens need to work for two objectives: the growth of the cable industry, and the channeling of that growth into imaginative and attractive forms of public service. These objectives will be hard to balance, because the people who want immediate profit from cable are not interested in developing its full potential for service. Therefore, strict regulation is needed to insure that part of the industry's profits be used to expand services to the community."[28]

There are, of course, options other than the franchising of a private entrepreneur--and the alternatives could well result in a more accountable cable management. James Richards cites *municipal ownership* as one possibility, in which a city, a board of education or a municipal department--a public utility commission, for example--might own the system. Acknowledging the dangers of political control and censorship inherent within municipal control, Richards sees an advantage in the profit potential it has for the city and its agencies. Yet another option is the ownership of a cable franchise by a *non-profit community corporation*. To quote Richards,

76

"The financial advantages to this form are that because profits are not being creamed, they can go right
back into the system to provide a number of features
which the other two forms cannot provide. Reduced
subscriber rates are one example. Certain free services
to members of deprived or poverty communities can be
afforded. Early sophistication of the system to provide
two-way audio and video signals can be introduced. The
cost of such sophistication is often the excuse given
by the private entrepreneur for not introducing them.[29]

Muncipally-owned cable systems are, in fact, being considered
by city councils in several metro communities. In a few instances,
local corporations have sought cable franchises for non-profit
systems. Several years ago, for example, New Samaritan Communications (whose parent structure is the Connecticut Conference of the
United Church of Christ), brought together representatives of
various Waterbury, Connecticut groups--business, church, educational and minority--to form a local corporation intent on offering
such advantages as we have suggested. New Samaritan has also
offered its assistance throughout the country to community groups
in seeking non-profit franchises. While the corporation has no
money to underwrite local groups, it does offer tax-sheltered
opportunities in assisting local coalitions to raise initial
study costs for a cable-system.

A third option to private cable ownership is a *cooperative
system* making the cable subscribers the organization's share-
holders. A citizens' panel in St. Paul, Minnesota brought to
its city council a recommendation recently for study of this
alternative--along with the former two. In addition, the panel's
recommendations included the establishment of a non-profit corporation to administer the three public access channels for which
the panel pleaded. Under this plan the cable operator would pay
the corporation a portion of its revenues, and the city would
contribute to the corporation's budget with funds it collects
in a franchise fee.[30]

Whatever the local administration of a cable system may be,
the congregations of a community--through representation on
citizen's panels, advisory committees, perhaps on the boards of
non-profit corporations--can be influential in assuring those
public service features earnestly sought by many in linkage
technology. As we have argued, this is a ministry in media
uniquely the province of the local church.

On the national denominational level, it is heartening to
report that enlightened church leaders have led the struggle to re-
dress some of broadcasting's abuses--racial and sexual discrimina-

tion, for example. One church media office which has had an effective ministry of critique to media industries and regulatory agencies is the Office of Communication of the United Church of Christ through the sophisticated direction of its head, Everett C. Parker. Focusing attention on existing broadcasting law and the Federal Communication Commission, the UCC agency has pressed the courts for humanistic legal interpretations, winning all its appeals and setting important precedent for public participation in the media, for more equitable employment practices and for fairness to ethnic minorities.[31]

From that office, Associate Director Ralph Jennings has recognized that the public needs "massive help" if it is ever to realize the full benefits of broadcasting--and that would include cable-TV. That help could come, he envisions, from Christian agencies.

"Churches are in a good position to provide it. They still reach relatively large numbers of people who are committed to Judeo-Christian ethical standards. They have a backlog of experience in sensitizing people to social and moral questions. This experience should now be used to help people to understand the crucial role of broadcasting in their lives and in our society. People can discover how to use television and radio creatively to enrich their existence. They can learn how to insist upon local broadcasting service which meets the community's needs and interests."[32]

Jennings' interest in local service would seem the special opportunity given us in linkage communication, and a special responsibility for the local congregation. Its ministry of critique could well encourage the development of cable's unique localism functions and sound the alarm when these are in jeopardy.

Noting, however, the all-too-common acquiescence of church leaders to the assumptions and values of media, Robert Louis Shayon believes

"the few voices among organized religious groups which have sought to change this pattern of accommodation to the broadcasters have not been widely heeded. The United Church of Christ is the only denomination which has involved itself, as a matter of policy, in bringing legal pressure to bear on the broadcasters. In so doing the Office of Communication

78

has interferred in the marketplace, and the church
has seldom been welcome there.[33]

But, lack of welcome has not deterred the UCC--nor indeed those
in church and public groups emboldened by its modeling of this media
ministry. In December, 1979, the Office of Communication's 15-year
struggle to protect the rights of blacks in Jackson, Mississippi
to television service was successfully ended with the award of the
WLBT-TV license to a black-controlled group. In recent months, the
office has parented the Telecommunication Consumer Coalition, made
up of some 200 church and public agencies nationwide. TCC--through
its members--has struggled against and so prevented congressional
and FCC efforts to de-control electronic media and to drop the
important principle of *public service* from communication law. It
has itself presented a full-blown re-write of the Communication
Act of 1934, has led in the study of cable and satellite technologies and in perceiving their implications for both nation and
world. The UCC ministry of critique has moved some mountains!

Local churches eager to have clarified their ministry to cable
regulatory agencies would do well to contact the Office of Communication, the Telecommunication Consumer Coalition or their own
denominational communication office. The National Council of
Churches' Communication Commission, the World Association of
Christian Communication (North American Broadcast Section) as well
as the Division for Film and Broadcasting of the U.S. Catholic
Conference provide other ecumenical sources of cable counsel.[34]

The Ministry of Communication

Major church denominations, over the years, have been involved
in few significant ministries of communication via mass media. To
be sure, today's *Electronic Church*, convened by so-called evangelical
Christians, summons millions and garners millions--in radio-TV audience numbers and contributed dollars![35] But, main-line churches,
Catholic and Protestant, for the most part have refused to make
direct over-the-air appeals for contributions, while attempting--
and failing--to claim significant hours of the broadcast day as
free public service time. The ministry of communication has been
feebly financed and largely ineffective. While there are notable
exceptions, few program efforts deserve plaudits and few have
managed lengthy tenure on the air. Television and film production
have become prohibitively expensive and are virtually unavailable
to the church. And, while radio so less costly and more accessible,

79

too few creative audio series have been developed. Rarely have the electric media been employed to probe the deep theological and ethical issues.

Again, the ministries of communication via mass media have hardly been possible (in important, continuing ways) to those whose locus of service is the parish. With few exceptions, church programming has been limited to broadcasts of the Sunday worship, of devotional *sign-ons* and *sign-offs* and of half-hearted discussion programs broadcast in the public service "ghetto" hours. When given real opportunity, however, we believe pastors and parishioners are capable of communicating with impact and significance.[36]

That opportunity would seem to be here--in cable television. Linkage technology provides the chance to communicate--both within the parish and to the community--at low cost, with continuing opportunity, and in prime time. The multiple channels afforded by cable eliminate the intensely competitive situation of broadcast media which has tended to crowd out the minority voice, including that of the church. Narrow-casting (pointing communication toward a particular audience without concern about its size) makes it possible to reconceive our ministries of communication within the parish, and for the whole ecclesia to get involved.

We would stress the notion that communication via cable is a shared ministry, one in which laity and clergy have important roles. It is certainly not a do-it-yourself project of the minister. Rather, as we see it, the clergyperson has mainly an enabling, facilitating function--exciting interest in cable uses, locating opportunities, encouraging creativity. The technology provides a significant occasion for the priesthood of all believers to be expressed. It will be difficult for some clerics to function in this style. And, predictably, if given freedom in the matter, lay persons will say and do things in cable-casts that jar theological sensibilities. On the other hand, if the laity is trusted and ministers participate in the planning, over time we would look for increasing maturity to be demonstrated. We particularly suggest that *young people* be given opportunity in cable communication. They, after all, are the "TV children" and often understand the medium much better than their print-oriented elders.

Margaret Mead, when she herself was no longer young, made an important point in her discussion of changing patterns of education.[37] Her belief was that we have shifted from a culture that is "post-figurative" (one in which young learn from the old) to one that is "pre-figurative" a culture in which--as the future explodes into

the present--the old learn to learn from the young!   "Today," she
says,

> "everyone born and bred before W.W. II is an immigrant
> in time--as his [sic] forebears were in space-- strug-
> gling to grapple with the unfamiliar conditions of
> life in a new era. But young people are at home in
> this time. . . .They can see elders using means that
> are inappropriate and that their performance is poor
> and the outcome very uncertain. They don't always
> know what must be done, but they feel there must be a
> better-way."

Expanding her point elsewhere, and with special relevance to
mass media, Mead suggests that people in their middle and older years
find themselves strangers in a new world.

> "All of us who grew up before World War II are
> pioneers, immigrants in time who have left behind
> our familiar worlds to live in a new age under con-
> ditions that are different from any we have known.
> Our thinking still binds us to the past--to the
> world as it existed in our childhood and youth.
> Born and bred before the electronic revolution,
> most of us do not realize what it means."[38]

In learning the rudiments of cable-TV and developing expertise
in communication, we in the church would do well to adopt the pre-
figurative stance--giving young people (who have video *in their
bones*) the chance to lead and teach in the use of the new medium.[39]
Ministers may indeed have special skills for cable communication--
experienced as they are in public speaking and perhaps with some
sophisticated understanding of mass media. We do not mean to discourage an active role for clergy in the ministry of communication.
On the other hand, they need not be burdened with the major responsibility of developing a whole new program in the parish. Cable, as
we have said, can become a means to facilitate many of the ministries
already alive in the congregation and should not be seen as adding
to the busy-ness of already busy church professionals. Further,
while radio-television personalities have been cultivated in the
broadcasting industry, star performers seem less appropriate to a
community which emphasizes the value of every member and each person's contribution.

Church cable-communication, in fact, ought immediately be seen
in its wider possibilities for community involvement. In our view,

the ministry of communication demonstrates "ecumunity".[40] It is
ecumenical in the way it often reaches beyond the parish to involve other congregations, and beyond the churches to engage
issues in community groups both congenial and confrontive. Vital
dialogue obviously should reach to embrace all in the wired
community, giving a voice to those who have been denied one,
risking the clash of opinion and theological debate, providing an
electronic soap-box for the lone spokesperson and a forum for
significant sharing. We are suggesting that church leaders join
with others in summoning the community to a new kind of *Town
Meeting*,[41] one in which the church is there not to proselytize
but to join in honest partnership with those whose interest is
community betterment. Cable-television, we truly believe, gives
us a new mechanism for dialogue in the neighborhood, for achiev-
ing new understandings of persons and issues and for initiating
positive action on matters often neglected. We are convinced that
local congregations can be effective agencies in facilitating
communication flow in a community.

Again, it will be difficult for the church on occasion to dis-
dain advocacy for the role of dialogue-enabler. But in the latter
style perhaps, the Christian gospel is often articulated more con-
vincingly. In effect we shall be *telling* by our *doing*, to pick up
an emphasis of those re-thinking the meaning of evangelism.[42] A
built-in danger of public relations programs is that they tend to
be protective and promotional. But, as many would suggest, the
body of Christ ought to be less defensive of the institution and
more interested in tragic realities of a sick world. Institutional
self-interest, in fact, would seem to *incarnate irrelevance*, since
such self-interest does not serve the community. By contrast, the
Lord of the Church is believed to incarnate what is most real about
God and about humankind, *Relevance* with a capitol R! Perhaps we
need to recall that Jesus' most angry words were for the institutional church of his time and for religious professionals, both
of whom appeared to be inward-looking and self-preoccupied.

Archbishop William Temple once said that the church is the
only organization in the world that exists primarily for those
who are not its members. That may be a good reminder to us in our
ministry of communication, tempted as we shall be to utilize new
opportunities in dominantly promotional ways. Cable-TV may certainly be a means to bring more people to the church, and that is
much to be desired. More to our purpose, perhaps, linkage systems
may enable us to bring more of the church to the world.[43] In that
enterprise we look to cable-TV's access opportunities with hope.

Our optimism would be misplaced, however, if we did not also

see churches and other public groups acting to assure the availability of access channels, and helpfully using them when they are
in place. For, in truth, the story of public utilization of access
opportunities is largely one of apathy and neglect. In some ways,
groups which have engaged in cable experiment have simply not
appreciated the amount of effort needed to use the medium well.
They have been indecisive about what they wanted to communicate and
shoddy in their preparations. Some community production teams have
no doubt expected too much of their early efforts and been quickly
frustrated. Long-range objectives and adequate organization for
follow-through are clearly needed. Good first programs need to be
followed by seconds!

The same disappointment is appropriate to much of the church's
effort in the world of cable access. Still, it can be reported
that church cablecasts are occurring in many sections of the
country, some of which represent exciting exploration of the medium's
possibilities. United Methodist church leaders in several Indiana
communities, for example, have originated cable programs of religious
news, denominational and Bible history, youth issues and the church,
as well as a 12-week teacher's training series. Senior citizens,
high school youth and community clergy have been encouraged to
develop in-church videotapes and to increase their video literacy.
A similar interest, using a related technology, inspired eight
churches in a Philadelphia presbytery to help young people use
biblical motifs in probing local issues on Super-8mm motion picture film.[44]

Elsewhere, Episcopal churches in Knoxville, Tennessee and
Richmond, Virginia have become deeply involved with local origination via cable, locating funds for video equipment and program production and once again becoming a center of the community's communication.[45] Another church functioning as a communication resource
to the whole community is Trinity-on-the-Hill United Methodist
Church in Augusta, Georgia. In 1975, David Pomeroy reported,

> "Trinity has, in effect become the production center
> for origination Channel 5 of Augusta's Cablevision
> system. . . .While there is a fair amount of expli-
> citly religious programming. . . Trinity continues to
> emphasize that it wants do do--and does--programming
> from all aspects of the community."[46]

Other local church groups have sought ecumenical coalitions
to enable their cable communication. In Madison, Wisconsin, the
Madison Area Community of Churches and the Roman Catholic Diocese

several years ago began production of a series titled "The Ultimate Fuel," with a magazine-style format and on-location videotaping. Its design was to "increase the awareness of religiously-motivated activities" in the metro community. Meanwhile, the Ministerial association of Bloomer--a small town in northern Wisconsin--wakened a community to its access opportunity, created the first program of local origination and served as a community catalyst in exploiting a medium which had served the town for many years but had never provided local video.[47]

We cannot catalog an exhaustive list of church groups involved in cable communication, though we should mention those in Long Beach, California, in Buffalo and Kingston, New York, in North Bergen, New Jersey and in El Paso, Texas who have pioneered in local origination.

Still other congregations have leased a cable channel for their programming. Grace Presbyterian Church in Peoria, Illinois, for example, with a history of radio broadcasting, wired its sanctuary for lights and cameras and obtained its own channel when the local cablevision system was ready.[48] Groups in Castro Valley, California, Tulsa and St. Louis have shown interest in similar leased arrangements. Accompanying this local activity, there has been noteworthy regional and national church interest. Several church council broadcasters, denominational communication officials and representatives of the National Council of Churches Communication Commission have provided helpful models, kits and counsel to congregations and coalitions wanting to expand their ministry of communication into cable-TV. In Canada, where cable development is perhaps ahead of the United States, the Interchurch Broadcasting National Study Committee of Cable and Community has issues a book-length collection of program formats for local origination and for cable workshops.

In the chapter which follows, we ourselves provide models for cablecasting, some of which are designed particularly to stress *ecumunity*. Our discussion of church and community communication also includes some strategies for more significant parish involvement in the beckoning world of cable.

NOTES

[1]*Broadcasting Yearbook* in 1971 cited six hours, 18 minutes a day as the amount of time given television viewing by the average American family. Ben H. Bagdikian, in "The Narrow-Gauge, Broad-Band Medium," *et cetera* (June, 1977), p. 204, indicated the

viewing time had increased in the interim to six hours and 56
minutes a day.

2Robert Konzelman, "Swim in Their Waters If You Want to Save
Them," in *Learning with Adults in the Parish* July-Aug., 1973, p. 28.

3Pierre Babin, ed., *The Audio-Visual Man* (Dayton, Ohio: Pflaum,
Pub., 1970).

4Quite obviously, the statement was *not* heard at United Theological
Seminary of the Twin Cities, where concern for media ministries
has long been encouraged by both administration and faculty.

5*The Television-Radio Audience and Religion* (New York: Harper and
Bros., 1955), p. 403.

6"TV and the Church: A Positive Stance," in *Arts in Context* (April,
1977), p. 1.

7"The Dust of Death" (London: Intervaristy Press, 1973), p. 374.

8"What TV is Really Doing to Your Children," in *Family Circle*
(March, 1967).

9The term "popcorn syndrome" and its definition come from Thelma
Altshuler and Richard Paul Janare in *Response to Drama: An Intro-
duction to Plays and Movies* (Boston: Houghton-Mifflin, 1967).

10"Psychopathology of Television Watching," in *Intellectual Digest*
(Nov., 1972), p. 26.

11For a selective index of such material's now rather dated, see
Robert W. Clyde and Eugene C. Jaberg, "The Use of Mass Media in
Religiously Motivated Adult Education: A Review of the Literature,"
in *Mass Media/Adult Education* (Columbus, Ohio: Ohio State University, Feb., 1971).

12*Media and Methods* (Circulation Office at 401 North Broad Street,
Philadelphia, PA, 19108) regularly publishes "Prime Time School
Television," including a calendar of upcoming TV specials and
several-page guides to note-worthy features. While designed for
secondary school use, the guides can readily be adapted for adult

church groups. Other helpful listings and discussion aids for television and film can be found in such publications as Mass Media Newsletter (2116 North Charles Street, Baltimore, MD, 21218) and *Media Mix* (221 West Madison Street, Chicago, IL, 60606). Denominational publications are often another source of helpful guides.

[13]For a development of this notion, see William Stephenson, *The Play Theory of Mass Communication* (Chicago: University of Chicago Press, 1967).

[14]Among those who have pioneered in consumer education for children are Shirley and Dean Lieberman of St. Paul, Minnesota, who--for many years--have worked at developing curriculum materials at various age levels designed to teach critical viewing skills. The Liebermans have established the *Institute for Visual Learning, Inc.* (1061 Brooks Avenue, St. Paul, MN, 55113) published *Project Focus Viewsletter* and are in demand as speakers on teaching children about television.

[15]Quoted by Kate Moody, "Many Schools Now Trying to Help Children View TV Critically," in a *New York Times Service* feature, *The Minneapolis Tribune* (Jan. 13, 1980), p. 14f.

[16]*Ibid.*

[17]"What Cable Can Do For You," in A.*D.* (May, 1973), pp. 38-39.

[18]*Ibid.*, p. 39.

[19]*Morality and the Mass Media* (Nashville: Broadman Press, 1968), p. 118.

[20]The reference is to the *Radio Act* of 1927 and the *Communication Act* of 1934, the latter establishing the Federal Communication Commission. The Communication Act, now 46 years old, certainly needs re-writing, but recent congressional efforts to do so have been met with alarm by church and public groups who see the *public service* philosophy of broadcasting in jeopardy. Congress seems intent on "de-regulating" electronic media, and that also appears to be FCC sentiment.

[21]*The Church in the World of Radio-Television* (New York: Association Press, 1960), pp. 55-56. For stylistic purposes we have used a paraphrase very similar to Bachman's statement.

[22]*Op. cit.*, pp. 31ff.

[23]*Ibid.*, p. 84.

[24]*The Times* (London, April 21, 1975), p. 2.

[25]We recall an instance in which no more than a dozen letters convinced television network officials to restore to a local station a program they had prohibited the channel to telecast.

[26]Ultimately, we might ask, ought not cable service be given free of charge (in a manner analagous to other welfare provisions) to those unable to afford it?

[27]A "position paper" detailing possible provisions for a cable franchise was offered in testimony to the Minneapolis, Minnesota Cable Television Committee, on August 17, 1972, by Patricia McKay, representing the Minnesota Council of Churches and the Rev. Robert Nygaard, representing the Office of Communication, Catholic Archdiocese of St. Paul. We have used the paper as a resource for this discussion.

[28]Everett C. Parker, "Cable-TV--How it Began, Where It's Going," in A.*D.*, (May, 1973), p. 38.

[29]James Richards, *op. cit.*, p. 5.

[30]"St. Paul Gets Suggestions on Cable-TV," in *The Minneapolis Tribune* (Nov. 8, 1979), p. 17B.

[31]Robert Lewis Shayon, "Fairness on the Air: How the United Church of Christ Has Helped Improve Radio and TV," in A.*D.* (Sept., 1972), p. 66. Shayon lauds, in particular, the ministry of Everett C. Parker, long-time Director of the UCC Office of Communication. More recently, Parker was the lone church leader named by *Broadcasting* (Jan. 7, 1980), p. 82, among its "Fifty-four Men and Women Who Put Their Media Mark on the 1970's."

[32]"The Gospel According to TV," in *Engage/Social Action* (June, 1976).

[33]*Op. cit.*, p. 68.

[34]Addresses of the agencies mentioned are these: The Office of Communication, UCC, 105 Madison Ave., New York, N.Y., 10016; The Telecommunications Consumer Coalition, in care of Janice Engsberg, Field Director, same address as above; National Council of Churches Communication Commission, 475 Riverside Drive, New York, N.Y., 10027; The World Association for Christian Communication, North American Regional Association, in care of John Magnum (chairperson), 231 Madison Ave., New York, N.Y., 10016; and The Division for Film and Broadcasting of the U. S. Catholic Conference, 13th Floor, 1011 First Avenue, New York, N.Y., 10022.

[35]On the other hand, there is little evidence that such communication functions as it purports to do, i.e. that it converts outsiders to faith. Mass communication seems far more likely to *reinforce* opinion than to *change* it, as indicated by much research. For a helpful summary of so-called media effects, see Wilbur Schramm, *Men, Messages and Media* (New York: Harper and Row, Pub., 1973), p. 241.

[36]Students of ours, equipped with some basic understandings of media and some production experience have often found uncommon broadcasting opportunities in their ministry. One young minister logged over forty radio programs, offered as public service time, in his first year in a Midwestern town and country parish.

[37]*Culture and Commitment: A Study of the Generation Gap* (Garden City, N.Y.: Doubleday and Co., Inc., 1970), p. 72.

[38]*Ibid.*, p. 74.

[39]Our experience in teaching media use and communication issues to seminarians continues to reveal remarkable understandings and creative uses of the media. It is clear to those who "instruct" in such circumstances that they had better be open also to instruction!

[40]"Ecumunity" is a term coined by C. Edward Johnson in 1967, when he was Public Relations Director for the Greater Minneapolis

Council of Churches. See *The Minneapolis Star* (Nov. 6, 1967),
p. 8A.

[41]"Town Meeting" was, in fact, the title given to a month-long
series of events in the Twin Cities of St. Paul and Minneapolis,
MN, in November of 1966. Community discussions, informed by media
presentations on local issues, and staged in churches, schools
and service clubs, served to focus attention on a variety of
metropolitan problems. And, the church-generated town meeting
spawned similar discussion-events in several other cities. For
a descriptive-evaluative discussion of this, see Eugene C.
Jaberg's "The Town Meeting of the Twin Cities: A Case History
of Community Dialogue," unpublished Doctor of Philosophy Disser-
tation, University of Wisconsin, 1968.

[42]An excellent small book developing this notion is Gabriel Fackre's
Do and Tell: Engagement Evangelism in the '70's, (Grand Rapids:
Eerdmans, 1973).

[43]An important discussion of church public relations activity that
is (somewhat uniquely) theologically substantive, is James E.
Sellers' *The Outsider and the Word of God,* (New York: Abingdon
Press, 1961).

[44]*A.D.* (November, 1974), p. 36.

[45]Information here is gleaned from a "short course" titled "The
Wired Parish: An Update on Church and Community Use of Cable-
TV," conducted by George Conklin at the Speech Communication
Association convention in San Francisco, Dec. 27, 1977.

[46]David Pomeroy, "Churches and Cable," an unpublished article
distributed by the National Council of Churches Communication
Commission in New York, 1977, p. 1.

[47]The fascinating story of local television in this tiny Wisconsin
community was told in an article headlined "Bloomer Likes Local
Television," in the *St. Paul Sunday Pioneer Press* (Jan. 26, 1975),
pp. 1, 4.

[48]"Peoria Congregation Leases Local Cable-Tv Channel," in *A.D.*
(Nov., 1974), p. 37

[49]David Pomeroy's article mentioned above as well as "Cable-TV and Video Resources List" and other occasional papers of recent date distributed by the National Council of Churches Communication Commission are helpful compilations of cable resources and available free in individual copies from 475 Riverside Drive, New York, N.Y., 10027.

"We must all become TV producers, knowledgable with camera and tape."

Robert Lewis Shayon, *The Crowd Catchers*

CHAPTER FIVE

MODELS AND STRATEGIES

With Chuck Anderson, we believe in "people making their own television."[1] That is a bright hope, the prospect of the *video pencil*, with the technology of two-way cable giving all of us a basic but supremely expressive communication tool. But the video pencil could be snatched from our grasp if we do not insist on local origination access programming as a public right and a vital aspect of cable-TV development, and if we do not develop some viable program models for responsible use of the access opportunities. The parish church--in partnership with other churches and the public institutions in its neighborhood--should be looked on to provide the ecumunity leadership to insure *local O*, the mechanism which might well "turn the medium of television around." With Anderson, we see a uniquely promising communication means now available as a "hands-on" experience to virtually everyone, and we want to encourage people to reach for it.

> "In a democratic society, active dialogue is held
> to be the ideal approach to problem-solving. There
> was once a time when a broad representation of the
> community was able to get together in the town hall
> and hold this kind of dialogue. People felt respon-
> sible and actively involved in the decision-making
> process. They felt they had some control over their
> lives. In the last hundred years, the social unit
> known as the 'community' has changed. Cities have be-
> come city-states. Many small towns have become voice-
> less neighborhoods in a sprawling complex of suburbs
> that stretch from one large city to another. One of
> the axioms of the twentieth century seems to be: when
> the size of a social group increases to the point where
> the individual member feels little or no participation
> in problem-solving, then community participation and
> spirit will diminish....In short, many of us have become
> apathetic because we feel helpless to influence the
> decisions that affect our lives....By using the television
> set that is in everyone's living room as a forum for
> community self-expression, we may be able to revitalize
> the democratic dialogue."[2]

From a church perspective, the enabling of the dialogue--in the community, in the congregation--is a prime objective of ministry. True dialogue--reciprocal communication, sharing of information, ideas and beliefs--is what uniquely fulfills and heals us. It is, as we have argued, the occasion when we touch significance, the moment bringing us intimations of the Divine. Cable-TV, then, with its two-way capacity, beckons as a special opportunity to promote the human dialogue. And for that special purpose, we suggest here some models and strategies for both church and community.

We see *models* of two kinds, one described as "presentation" while the other is more aptly termed "presence." *Cable presentations* would designate video experiences similar to those offered now by broadcast television. They would present church and community-designed video productions using the medium in its one-way capability, in which the viewer is a rather passive consumer. The *cable presence* of the church and community, by contrast, employs the two-way potentiality of the technology. The church or community agency, through its professional staff and through its members, becomes *present* in the places of reception because the dialogue which is the feature of face-to-face events now is possible. The experience is "live" and in real time. Cable *presence,* an increasingly available opportunity, will bourgeon, we believe, if there is strong public demand for its development.

Models for Cable Presentation

We do not construct our models as the last word for cable-casters. They are neither exhaustive nor definitive. We offer them, rather, to pique the imagination of the would-be producer for cable and to encourage exploration into some untried regions of creativity. The best program formats in a narrow-casting medium, will be those designed for particular audiences by those who understand the contexts and approach them knowingly. A single word of warning should be sounded. The initial efforts of in-dividuals and institutions to utilize cable may--by the "glossy" standards of current broadcast media--appear crude and amateurish. Thus, early experiment can lead to disappointment, discouragement and ultimately to abandonment of the effort altogether. This pit-fall can be avoided by developing a new standard of excellence, a standard which measures the worth of a production in terms of its relevance to a particular audience at which it is aimed, rather than upon its slick appearnce. The video pencil may write a shaky line at the beginning, but reasonable technical competence will come with practice and repeated experience. Meanwhile, let our focus be upon the quality of program content and the importance

94

of that content for the persons who will view it, and those who create it.

Video Information

One type of cable presentation provides *video information*--news of the community, taped documentary treatment of local issues, interviews of community people or of interesting visitors, perhaps a magazine style collage of informative features the focus being their relationship to the wired community. Such reports of the community might well be accompanied by a facsimile print-out also transmitted by cable and ready for the viewer to tear off the machine attached to the TV set. The obvious advantage of such narrow-casting is that people living in towns too small for local television (and that sould also include some very large suburban populations) now are given the chance truly to be in touch with their community.

A particular opportunity in this genre, would be the *video newsletter*--appropriate for churches as well as other public organizations. All institutions have information they must get out to their constituencies and to the community-at-large. Mundane some of it may be, but it is important to the group's functioning. Meeting times and places, special program opportunities which are available, needs to be met--these are a few of the bits of data organizations seek to communicate. In a real sense, the existence of the institution depends on how well such information is communicated. The mimeograph machine has been the mainstay of this kind of internal communication, but now cable provides a supplement to print!

What would a cable newsletter look like? Muse with us about this possibility: First Church, Anytown decides on this electronic means to reach its people, and contracts with the cable system for a 15-minute slot on one of the channels available for lease. The congregation is alerted to the event and invited to tune in for the weekly potpourri of information. The usual meetings, special events and educational opportunities are announced, perhaps enhanced by art work and slides. Then, some new faces appear on the screen! The minister or lay-person cable-caster introduces the members of a family which has recently joined the congregation and launches into an interview. Video creates an instant link between the face and name and the interview provides some helpful information about the people. Foundations for more personal relationships with others in the congregation are established, as the new family tells of its members' interests, hobbies, talents, travels, work and faith experiences. Persons who view are encouraged to pursue commonalities they might otherwise never have known about and people in larger congregations, over time, may really get to know their fellow members.

First Church also wishes to promote a special offering it will receive for the victims of a devastating flood in a neighboring community, or of a crop-failure and famine overseas. The staff decides to show videotape picturing the devastation on the cable system. The films, taken by a high school filmmaker in some cases or perhaps supplied by a denominational or ecumenical communication office, are accompanied by narration explaining what the church plans to do to help its neighbors. It may also want to do a follow-up story some weeks later indicating what is being done with the money given by First Church members. Whatever the eloquence of a verbal report from the pulpit, it could hardly equal the impact of this kind of video appeal. Two-way cable would allow this kind of presentation also to move toward video *presence*, with home viewers raising their questions and responding to the appeal with their concerns and their pledges of money.

We do not suggest that an electronic newsletter would replace the church's printed communications. One form, in fact, could be used to enhance the other. Printed pieces could focus on what that medium does best--expansion on ideas and development of issues in depth. Cable could be used to introduce new programs, new ideas and new directions (as well as new faces) while print communication could provide important background information and suggest the means of implementation--material which requires reflection and some digesting. The final content of the video newsletter, of course, emerges from its specific institutional context and would be limited only by the imagination of those who produce it.

Video Education

An intriguing form of the video presentation is cable *education*, for the technology delivering high-fidelity sound and close-up pictures seems to afford both an intense and highly personalized possibility for learning. Even without the two-way opportunity, teacher and students could become pedagogical partners, as it were, in ways that can happen only rarely in the classroom. For, video provides a dramatic visual "experience" of the material for learn-ing and of the person providing the instruction. There is much evidence to suggest that viewers react very personally to television celebrities, with whom they often feel closely acquainted. When the student has the added advantage of "up-stream capability," the educational experience promises to increase in richness.

In our schools systems, small-format video has long since established itself as an important adjunct to education. Everywhere schools use television equipment--picking up broadcast signals for classroom viewing, using pre-taped instructional materials, or creating their own student-produced programs. Churches, however, have lagged far behind. Despite the fact that most congregations give their educational ministry high priority, with a few notable exceptions video use is practically non-existent in the parish church school.

But, to return to our mythical congregation gathered at First Church, let us ponder what opportunities cable presents for its educational mission. Suppose the church wants teachers in its church school to participate in a training workshop. Often, in its experience, it has been very difficult to get its staff of volunteer teachers to attend events beyond what is considered the normal life of the congregation. If the event has been out-of-town and will require any significant investment of time, the recruiting task has been almost impossible. Now, however, First Church has access to a teacher-training workshop, the major components of which have been videotaped at a regional theological seminary. With accompanying print materials such as a workbook and assignment sheets, the presentation is laid on the local cable system and teachers participate without having to leave their homes. The church might join with other Anytown congregations in sponsoring the training event ecumenically. Church school teachers would have a growth-experience and at the same time ecumenical relations in the community would be fostered.

And, what of continuing education programs for clergy and other professional church workers? The specialized nature of such programs has made the use of broadcast media prohibitively expensive, assuming that there would be broadcast time available under any circumstance. On a cable system, however, through narrowcasting, special educational opportunities become feasible in a new way. In non-metropolitan areas where the resources of educational institutions are distant, vital programs of continuing education can now be offered--programs of the same quality and content as those available from a university or seminary. Occasional discussion seminars could be staged at central locations and systems of testing and grading for graduate credit could be devised, much as some schools currently offer home TV courses for credit. The *Open University*, which has functioned successfully in the United Kingdom for many years, becomes a particular model for this kind of media implemented higher education. Open University *graduates*--for it is possible to earn a bachelor's degree through this process--are everywhere in the British professions and, in our judgment, often evidence superior educational preparation for their work.

There are yet other possibilities in cable-enabled education. Family Bible study courses could be designed to utilize the family unit as a learning environment. Families--whatever their pattern--could be urged to watch the regularly-scheduled series, and follow the cabled discussion with one of their own, using questions suggested on the program as their guide. A supplemental program might be developed for Sundays in which participating families and single persons could come together, discuss their learning, exchange ideas and insights and ask questions about issues that are unresolved.

If there is a particular area of the church's educational ministry in which video could be used extensively it would be adult education. As someone has well observed, Jesus taught the adults and sent the children off to play. In church education we have often reversed the process. Video presentations dealing with issues in medical ethics, theology and biblical research could enhance adult learning in virtually every parish in the land. Important questions of the faith could be aired and discussed. Denominational leaders whose names--much less faces--are often unknown to local congregations, would have the opportunity to address their constituency and (anticipating video presence, for a moment) often obtain immediate feedback from the grass roots about proposed programs and denominational directions. Such a person-to-person approach via the cable could serve to create a new sense of community in and among denominational bodies as well as imparting the kinds of information which now gets lost in the papers piled on the desktop of a beleagured pastor.

To re-iterate, the localized and localizable nature of cable makes possible programming for special groups within a congregation and community. And the re-assignment of funds within the church budget may put costs of cable production and/or leasing within the range of many groups. Hence, congregations could program specifically for the elderly, or youth, or handicapped persons such as the deaf. A production aimed at young people, for example, winsomely done with Christian substance and laid on the cable on a Saturday morning, would provide a welcomed alternative to the mindless cartoon palaver aired on commercial networks in the same time period.

Robert Konzelman, speculating on how linkage technology will affect the church, believes that the much talked about "house church" might become a practical reality in the cable age.[3] It could well be the locus for much of the educational opportunity we have envisioned, with neighbors meeting in one another's homes and *taking part* in church-originated cablecasts. Of course, that scenario would bring about some distinct changes to congregational life.

"Cable, in use by the church for a large part of its work, especially in education, would require different organizational structures and personnel. The effect of strong house churches or neighborhood groups would be to provide more and different types of leaders. Teachers, for instance, in the typical classroom sense might be fewer and discussion leaders and facilitators more numerous. New skills would be needed by members of the church for participation in such groups."[4]

As these alternatives are tested and as people get accustomed to the feel of the video pencil, the medium itself will likely become the subject of education. The encouragement to participate and to create, which is inherent in cable communication, will itself evolve into an educational experience. And, in the process of trying to say some things in new ways, persons may learn to listen more fully and sensitively than they have in the past.

Video Art Forms

Experiment with the most modest video equipment reveals the medium's potentiality in creating distinctive visual art effects. Aiming a camera at the TV monitor, for example, causes a kind of "video echo," with the image repeated in successively smaller replications to infinity. Rapid movement of the camera past light sources can create interesting streaks and swirls--this procedure runs the risk of damage to sensitive image tubes in the apparatus. In truth, a myriad of special effects are possible, many of which are distinctively those of the medium, and some of which will still be discovered in the video play of the amateur. Kas Kalba has labeled such experimental departure from TV conventions as "conceptual video," video happenings designed as artistic creations.

Other art forms can furnish the materials for cable presentations, some of which are enhanced by the peculiar properties of video. Drama, for example, is an art with a long history of significance to the church. The ecclesia finds itself called on to speak of a love which defies description, but for which poetry and drama are especially expressive, especially when given some new dimensions via electronic technology. Some things can be done with video which are simply impossible on stage. Special effects can be creatively utilized, indeed.

We include here an example of religious drama written specifically for video. The drama entitled "Bread" is based on the Lukan account (4:1-13) of Jesus' 40-day ordeal of temptation in the wilderness. A single actor was used to play both the Tempter and the Tempted, a theological statement in itself. The actor was videotaped playing one role, that of the Tempter. Then, utilizing the chroma-key capacity of television, he played the second role with his image superimposed on the first scene. The effect of the two persons in conversation is startling and intriguing.

Bread: A Modern Version of the Temptation of Jesus[5]

(We have chosen not to include specific video instructions here, allowing that dimension to be filled in by the reader's imagination. We put the Tempted figure on a stool, center stage, and had the Tempter circle around him, occasionally moving to approach the Tempted, speaking--as it were--over his shoulder. The actor played roles in the same costume, a black turtle-neck sweater and jeans).

Narrator   -   "And Jesus was led into the desert by the Spirit and after spending 40 days and nights without food, Jesus was hungry. And the tempter came to him and said, 'If you are the Son of God, turn these stones into bread...'"

Tempted   -   (EYES CLOSED AND TALKING TO HIMSELF). God, I'm hungry. (LICKS LIPS AS IF TO TASTE IMAGINARY FOOD). Oh God, I'm so very, very hungry.

Tempter   -   Of course you are.

Tempted   -   (OPENS EYES, STARTLED) What was...who was that?

Tempter   -   (CHUCKLING) Don't get excited. All I said was, "Of course you're hungry."

Tempted	-	(STANDING UP, LOOKING AROUND, TRYING TO IDENTIFY SOURCE OF VOICE.) Who are you?
Tempter	-	Does that really matter?
Tempted	-	Where are you?
Tempter	-	Does that matter either? Good grief, a couple of minutes ago you were hungry, now you want to play twenty questions. Look, are you hungry or aren't you?
Tempted	-	What does that matter to you? (WEARY OF HAVING TO CARRY ON DIALOGUE, RETURNS TO STOOL).
Tempter	-	There you go again! You know you should consider a career in religion. You have a marvelous knack for avoiding issues and dodging questions. You'd make a great theologian. I ask a simple question. Do I get an answer? No, I get three more questions.
Tempted	-	(NOW GETTING A LITTLE IRRITATED). All right! All right, I'm hungry. Yes, I'm hungry, does that make you happy to hear?
Tempter	-	No need to get huffy! I'm not condemning you for bring hungry. You've been out here a long time. Let's see, how long has it been?
Tempted	-	(HEAD BACK IN HIS HANDS). 35, 40 days. I don't know. I've lost track. Look, why don't you just go away!
Tempter	-	40 days, that's a long time. You can get mighty hungry in 40 days. (THEN, AS IF COMING UP WITH A GREAT IDEA). Look, why don't you go back?
Tempted	-	I can't.
Tempter	-	Its not that far.
Tempted	-	(A BIT MORE AGITATED). I can't.
Tempter	-	Look, I'll go with you. In a couple of hours we could have all the food we could possibly ever want.

Tempted   -   (NOW IN REAL AGONY). I can't.

Tempter   -   (AS IF UNAWARE OF RESPONSE). Fruits, meats, salads!
You name it and we'll have it. Every kind of imagi-
nable delicacy and we'll eat until we...

Tempted   -   (NOW SHOUTING). I can't!!

Tempter   -   You can't? Don't be a damned fool. What are you
trying to accomplish? What are you trying to prove?
Do you want to be some kind of martyr, some kind of
saint?

Tempted   -   (IN ANGUISH). I don't know. I don't know. For God's
sake, leave me alone.

Tempter   -   (PRESSING THE POINT HARD). The world doesn't want
saints and martyrs. The fact of the matter is they
can't stand them. Oh sure, they celebrate them, they
tell stories and sing songs, but they really can't
stand them. Why else do they ridicule and kill them?
Is that what you want--to be laughed at, to be killed
by those unfeeling, uncaring slobs back there?

Tempted   -   Please!

Tempter   -   Or maybe you'd like a footnote in the Guinness Book.
"The world's longest hunger strike. 40 days and
40 nights without food." The feat could take its
place alongside such monuments as the world's record
for standing on your head or the longest ping-pong
game.

Tempted   -   Please, please, just leave me alone.

Tempter   -   (CONTINUING TO PRESS). If you're lucky maybe someone might even think up a good reason for this silli-
ness. "He did it for mankind." "He gave himself for
others." Oh that's beautiful, but its not going to
fill your belly, brother!

Tempted   -   (SHOUTING NOW, HOLDING HANDS OVER EARS, TRYING TO
SHUT OUT THE VOICE). Stop it, stop it! For God's
sake, stop it!!!

Tempter	-	(NOW SUDDENLY TURNING CONCILIATORY). Look, I don't want to upset you. All I'm trying to do is help you see things in their proper perspective, the way they really are. Look, who really cares?
Tempted	-	(SHOUTING BACK). I care!
Tempter	-	Now what kind of answer is that?
Tempted	-	(RAISES HIS HEAD AND LOOKS AS IF TO TRY AND GRASP SOME KIND OF INSIGHT). No, no, wait a minute. I care, I really do. I just realized that. (GETS UP SLOWLY). Maybe that's why I'm out here. Maybe that's why I'm hungry. I care. I care!
Tempter	-	Well, it's really gotten to you, hasn't it? You've been out here by yourself too long. You're starting to lose your grip.
Tempted	-	No, I'm not. I think I'm just starting to get it. I'm hungry, right? (DOESN'T WAIT FOR AN ANSWER, BUT GOES ON, GETTING MORE EXCITED). Okay, I've got this pain in my guts but I'm still alive. I can live with that pain.
Tempter	-	And just what does that prove; that you're in better physical shape than most? Maybe you're just lucky.
Tempted	-	You don't understand, do you? If I can have this pain, if I can be hungry and still feel alive, it means there must be more to living than just eating. There must be something that renews and recreates and sustains me as a person.
Tempter	-	Brilliant insight! And now you can go back and tell people "Stop eating, folks, I've got something better"!! You are crazy with the heat. They're going to lock you up.
Tempted	-	No, I'm not saying that. Don't twist my words. I'm not saying "Don't eat." I'm saying there's more to life than that. Food can maintain existence but it can't sustain life; there's more. There's more here in me, between me and other people. That's what makes it and I've got to show them.

103

Tempter   -   Delusions of grandeur! I knew it. The heat, the
                isolation. YOU are going to show them. YOU? You
                are going to tell them there's more to life than
                life. YOU? Do you know what you are? You're a
                joke; a miserable, pathetic, deluded, misdirected
                joke!

Tempted   -   It's not a joke; this ache in my guts is no joke.
                And this ache is telling me something about myself,
                about what real living is all about. I'm aching
                but I'm alive and it's not the food, the goodies,
                the externals making me alive. It's something else.

Tempter   -   Bread, that's what it is all about, brother, and
                don't kid yourself. Try and peddle that line of
                yours to those people out there and they're going
                to laugh you out of town...or worse!

Tempted   -   No, they won't. Not when they hear. Not when they
                see.

Tempter   :   (LAUGHING DERISIVELY).

Tempted   -   What's so funny?

Tempter   -   "When they hear, when they see..." as if they're
                going to, as if they want to. How can they see
                when their eyes are swelled shut with fat? How
                can they hear when their ears are stuffed with
                promises of fame, riches, security, gratification.
                They don't want words, they want bread and they'll
                sell their souls for enough food to keep their
                bellies full. I know.

Tempted   -   Shut up!

Tempter   -   (SURPRISED BY THE BLUNTNESS). What did you say?

Tempted   -   I said, shut up.

Tempter   -   Can't take the truth, eh?

Tempted   -   I can't take your line that there's nothing out
                there but a bunch of grovelling, bread-begging
                animals. You're dead wrong. Sure, you can buy
                their souls with bread. But they're going to find

out bread doesn't stop the aching. It doesn't stop the pain of loneliness, the agony of being hated and the stupidity of destroying. I'm beginning to see. There are all kinds of hunger and the deepest aches are the kind bread can't touch. That's the pain I have to stop.

Tempter   -   You're a fool!

Tempted   -   Maybe.

Tempter   -   You want to be a fool?

Tempted   -   There are worse things, I guess.

Tempter   -   Like what, O wise one.

Tempted   -   Like being empty, empty in here (POINTING TO HEART). Like feeling that your life is a vacuum. Like never being able to feel the breeze run its fingers through your hair or being able to rip loose and sing the praises of a new day. Or never feeling warm tears running down your face or a smile work its way up from your heart onto your lips and reach across the room to other faces and other hearts. I can live with the kind of hunger I can feel in my stomach right now. But I can't live with those other kinds of hunger and I don't have to. I'm not going to.

Tempter   -   Okay, fool, have it your way. You sit here in your desert on your rock. You sit here and talk about breezes and tears and smiles and your "More to life." You'll see, brother, you will see.

Tempted   -   I guess I will. I guess we both will. And for some reason that doesn't bother me anymore. (RETURNS TO CHAIR).

Tempter   -   Are you sure you won't come back?

Tempted   -   No, it's not time yet.

Tempter   -   How about some bread, just a little bit?

Tempted   -   No thanks. I don't feel hungry anymore.

Narrator -   "And Jesus said, 'We [sic] cannot live by bread
                 alone.'"

END

(*Production note*: If the video production system does not have the
chromakey feature, one could play this scene with the Tempted figure
and the Tempter represented by a voice--the pre-recorded voice of
the same actor. The identical theological comment is made, and some
of the dramatic impact is preserved. Obviously, the voice track needs
to be carefully timed to allow appropriate space for the responses
in the dialogue).

The biblical narrative provides an almost inexhaustible amount
of material for such use of video. Theological sensitivity to the
meaning of a given text enables us to write material communicating
the Good News in ways which avoid, on the one hand, the deadening
confines of literalism and on the other the libertarian abuse of
biblical material. The relevant and engaging presentation of this
material is--in point of fact--one of the major missions to which
the church in every age has been called. We live in the video age
and the words of St. Paul take on new meaning and offer a fresh
imperative: "I become all things to all persons [sic] that by
all means I might save some."

Other forms of drama abound and furnish a wide array of resources for cable presentation. One series, designed for Lenten use
and employing the chancel as a stage, is based on Steve Allen's
Meeting of the Minds television program. We believe it readily
adaptable for use on cable television. Guy Stone, the series pro-
ducer-writer has called it "Real People,"[6] and cast people from
church and community in the roles of figures from history. Each
mid-week evening in Lent, two persons from the past (sometimes
the present, for some contemporaries were represented) appeared on
stage in appropriate costume and were invited by a moderator to
share their essential ideas. The evening then moved to dialogue
between the two-sometime chosen because they had strongly contrasting views or because of their interest in a common theme--and
finally to questions posed by the congregation. The actors needed
to immerse themselves, of course, into the role they played and
into the thought of their character.

For his experiment, Stone chose such duos as Simon Peter and
Satan (a Faustian Mephistophoeles), Martin Luther and the Apostle
Paul, Mary Magdelene and King David, Thomas the Apostle and Sister

Theresa, the Samaritan woman and Martin Luther King. "Real People" illustrates how original materials can be created out of the imagination of church people and we see it as promising adaptation for cable purposes.

Other forms of video art will emerge easily for church and community use. Local musical recitals, church choirs, school musical organizations, municipal bands--all promise cable presentations which will attract us because they are from our community, and we know the musicians! Displays of school painting and sculpture and photography as well as the shows of local artists could be nicely produced in video versions. Community film-makers and amateur video producers would have a made-to-order showcase for their work. There seems no limit to the opportunities to cablecast video art.

Video Worship

A fourth genre of the cable presentation, one belonging in a special way to the church, is that of *video worship*. Indeed, the use of electric mass media to share the worship experience has been the primary interest of church groups exploiting radio and television, sometimes to the exclusion of all else. The reason is obvious. Worship is the most fundamental and visible ecclesiastical activity. Even the most nominal church member, while aware of nothing else going on in the congregation, knows that on Sunday the community gathers for worship.

That gathering could well be altered, as we have intimated, when energy shortages discourage travel and cable technology enables our meeting electronically. We may well use cabled worship experiences in house-church groups as our regular opportunity to praise God and share the Good News. More occasional festival gatherings in the church sanctuary (or public building, adapted for the purpose) may continue to give us a sense of the total congregation and our fellowship in the wider communion of saints. Such total-congregation gatherings would also give opportunity to celebrate the sacraments in their usual style. Such a change in the worship of the parish church could well occur, without--or so it seems to us-- jeopardy to the essential experience of worship.

In the meantime, we see video not so much a vehicle of *replacement* but of *enhancement*. In addition to Sunday morning worship, churches could develop other devotional experiences designed

specifically for television, experiences which would utilize
visual images provided by art work, sculpture, slides and motion
pictures. Imagine for example a 15-minute video meditation based
on the theme of creation, or visuals used to illustrate a particular hymn or other piece of church music. Worship, we believe,
need not be limited to a particular time and space. If our
notion of God as concerned about and active in the world has any
validity, video presents the opportunity to muse on the images of
the world in ways which will make our theology even more real.

Images of the world--our world, our community, our church and
homes--were projected to a worshipping congregation at a regional
church meeting through a collection of carefully-edited slides.
The pictures were accompanied by a sound script (narration and
music), which explored some of the traditional metaphors for the
church. We present portions of the script as a model, believing
that it represents material useful in a video application. A cable
version of this sermonic piece would amount to more than a presentation of slides, of course, since the TV camera can move on the
image, zooming in to single out important detail, or panning and
tilting in movement over the surface of the picture to create
motion and scan the visual information. Here, then, without the
video column of instructions (details to be supplied by the cable
production team in light of their choice of slides and how to
present them) are selections from

Images: An Exploration of Meaning and Mission[7]

(ESTABLISH ORGAN THEME TEN SECONDS, THEN FADE UNDER FOR
NARRATION).

The church, that is what we represent here this morning.
But what is the church? Is it a building? Is it just
the individual congregations we represent and to which
we will return when this meeting is over today? Is it
just a voluntary organization of like kindred who
occupy a certain place at a certain time every week?
Or is it more.....more.....what more....???

(MUSIC UP FIVE SECONDS, THEN UNDER).

"You are a chosen race, a royal priesthood, a holy
nation, God's own people, called to proclaim the
wonderful acts of God, who called you out of darkness

into his own marvelous light. For once you were no
people, now you are God's people..."

People of God...one of the images of the church
found in the Bible, images which call us again and
again to review and reform our perceptions and understandings of this thing we call the church.

People of God, not persons, but a people, a community,
a nation of people bound together not by covenant or
contract but a call, the call of God--the call of the
Almighty, Creator God to community and peoplehood.

Community of the forgiven!

Community of the Redeemed!

Community of the saved!

Saved? Yes saved, but not so much saved *from* as saved
for! Saved as a community not to occupy a place of
privilege but saved for response and responsibility....

People of God...working people, servant people,
risking people...set aside but not set above. People
of God responsive to the cries of the hungry, the
lonely, the sick and oppressed.

"Once you were no people, now you are my people...
People of God."

(MUSIC UP FIVE SECONDS, THEN UNDER).

If the image of the church as people of God conveys
unity and a sense of community, the Biblical image
of the church as *Body of Christ* underlines and affirms
the diversity which exists among us.

"There are many parts but one body."

Different, yes but nevertheless parts of the single
body, body of Christ. And it is the oneness of that
body and the meadship of Jesus Christ which holds us
together in our differences, allows us to be many but
one, allows us to celebrate our differences as we
stand together....

For in the end, the image of the Body of Christ tells us our unity is dependent not upon dogma or doctrine, preachment or practice, law or liturgy. Our unity comes from Christ! Different and yet one, one in the spirit, one in the Lord, One Lord, one faith, one birth. The Body of Christ.

(MUSIC UP FIVE SECONDS, THEN UNDER).

"Now if anyone is in Christ he is a new creation; the old has passed away, behold the new has come."

The Biblical image of the church as part of a *new creation* opens fresh horizons for understanding the meaning and message and mission of the gathered community we call the church.

"New creation," writes the apostle Paul. Not the old revisited, reworked, revamped or reinterpreted. The work and Word of God in Jesus Christ opens a *new* age.

Age of renewal...renewal of the knowledge of God in those who choose righteousness, holiness and justice as a way of life.

Age when the distinctions of race, religion, culture and class will be transcended and overcome and as new solidarity emerge based upon the image of God, an image toward which the renewal of the age causes us to move....

And the Church is called to be part of that creation. Not just part, but a manifestation of that new creation here and now, a community of renewal in which the light of Jesus Christ has broken and begins to shine forth in the lives of persons and in the work of the community. The radical, risking community, willing to die in order that there be new life! For that is the way of the Kingdom, Kingdom of God, the *new creation*.

(MUSIC UP FIVE SECONDS, THEN UNDER).

Yet a fourth Biblical image of the church is the image of the *boat*. It was from a boat that, on several occasions, Jesus taught the people who had come to see and hear him. And it was to the disciples in a boat

that Jesus came in a raging storm and showed them
the miracle of faith.

It is not surprising that a boat should emerge as an
image of the church. After all it is a vehicle of
transport and we in the church are a traveling people,
a pilgrim people, called to venture out onto the raging
seas in the faith and trust that God will guide us.

The boat, a call to adventure in the company of fellow
adventurers.

For we do not travel alone. This is our affirmation
and proclamation. Our commission, our traveling papers
come from God.

"God calls us to share the cost and joy of discipleship."

(MUSIC UP THREE SECONDS, THEN UNDER).

For we are community you and I, a community whose history
and roots reach back to the prophets and apostles, back
to the martyrs and reformers. Like them we are called to
be the Church in our time as they were in theirs. Called
together, to stand together, pray, praise and serve
together, as

    "the people of God"

       "the body of Christ"

          "the new creation"

            "the ship"

Called by God in the name and for the sake of our Lord
Jesus Christ. Unto God be all blessing and honor, power
and glory, now and forever, world without end.

The visual elements of this sermon in its video form would,
then, be slides, some of which could be gleaned from the collections
of photographers in the congregation. Other slides, depicting some
of the specific requirements of the metaphorical themes, would need
to be created fresh for the presentation, either from prepared art
work or from staged scenes for photographing. The talent to achieve
this kind of video exists within many congregations, or--if not to

be found in the church--could fairly certainly be recruited from the community.

Another model for video worship also relies on slides for visual effect, and at the same time uses staged studio vignettes. The model here[8] is a program which was staged as an experimental worship experience for broadcast television, and is given in the usual two-column layout of the TV script.

In the Beginning: A TV Worship Service

VIDEO	AUDIO
FADE UP ON SLIDE: 1 (LS OF WORLD)	(MUSIC CREATION THEME UP, HOLD FIVE SECONDS, FADE UNDER NARRATION).
DISSOLVE TO SLIDE: 2 (CLOUD MASS)	NARRATOR: In the beginning.
	In the beginning, God!
	In the beginning, God created.
DISSOLVE TO SLIDE: 3 (GALAXY)	Heaven and earth, earth and heaven. All that was, is and shall be, God created.
DISSOLVE TO SLIDE: 4 (SUNRISE)	Let there be light!
	And there was.
	Let the land and the waters be separated!
	And they were.
(SLIDES CONTINUE THROUGH THIS SECTION)	Let the earth teem with living things; birds in the air, fish in the sea, animals of every kind abroad in the land. Let there be LIFE! Be fruitful and multiply.
	Multiply!
	Multiply!
	Multiply!!!

VIDEO	AUDIO

Man and woman, woman and man. They
too were there in the beginning.
Created in the image.

Image of who?

Image of God.

In the beginning.

And it was good; so very, very good.

All of this in the beginning and all
so very, very good. So good........
so good to begin.

VIDEO:

CUT TO MS
   (MAN IN BATHROBE)

CUT TO MCU PROFILES
   (TWO PERSONS FACING
    EACH OTHER)

AUDIO:

MAN: (LOOKING VERY QUIZZICAL). I
don't get that!

PERSON 1: Let's start.

PERSON 2: No, let's begin.

PERSON 1: It's all the same; beginning and starts, starts and beginnings. What's the difference?

PERSON 2: Difference? A great
difference!

PERSON 1: No difference, all the
same.

PERSON 2: Not the same. I start my
car, I don't begin it.

PERSON 1: Word games! Word games!
Just so many words games. Childish,
childish, petty absurd games.

PERSON 2: No, no, not so! This is
about beginnings, not starts. It's
new; *new*, not just redo. Begin, like

113

| VIDEO | AUDIO |

God began. With nothing, out of nothing. Fresh and clean; all pristine. That's "begin."

PERSON 1: Begin?

PERSON 2: Begin!

BOTH: (TOGETHER) Begin!

CUT TO LS
    ("PREACHER" FIGURE)

PREACHER: (Looks very quizzical) I don't get it either!

DISSOLVE TO SLIDES
    (THEY CONTINUE AT
    RAPID PACE THROUGH
    THIS SECTION)

(MUSIC: UP THREE SECONDS, FADE UNDER).

NARRATOR: Beginning, opening of life. A joyful sound, that cry. "I am here!" it says. Here, to occupy my space; that unique niche in all the universe which is mine and reserved for me. I am here to grasp the world and clutch it to me like the priceless gift it is. The sights, the sounds, the smells, the tastes. Everything is new to me and I love it!

Beginning, discovery that there are other people in the world with me, that I can share with them and they with me. My world expands, my horizons stretch out. School, church, neighborhood-- new worlds to explore. Everything I touch, everything I see, every place I go--I begin.

Beginning, finding there are special ones with whom to share in special ways. That almost imperceptible movement from acquaintance, to friend, to companion, to lover. Special ways with special people. Beginnings of new dimensions to life. Beginnings of love.

114

VIDEO	AUDIO

(SLIDES CONTINUE)

Beginning, and the further discovery there is indeed a special place for me, unique contributions I can make. Store, office, factory, home; the location matters not. It's my place, my work, my gift and I begin.

Beginnings for all of us. Start of life, discovery of family, job, self. New horizons, experiences. In those beginnings it all seems so beautiful, so hopeful, so endless. But as life goes on the beginnings seem to come further and further apart. Day in and day out; same old thing. A twist here, a turn but the same old thing. No more beginnings...Now come the ends. They have other names..."end" seems so final. But ends they are; end of family, end of job, end of life. No more beginnings....now only ends.

CUT TO MLS
   (MAN IN BATHROBE)

MAN: "I get that!"

DISSOLVE TO SLIDES
   (CONTINUING)

NARRATOR: New beginnings! Less dramatic perhaps, but no less significant.

"Truly I say to you unless one is born anew, he cannot see the Kingdom of God."

Born anew. That means a beginning, something new, something unique.

"But Nicodemus answered Jesus, 'How can a person be born again when they are old?'"

Good question! Let's see. Let's see if there are such things as new beginnings.

115

VIDEO	AUDIO

VIDEO

(SLIDES CONTINUED)

AUDIO

NARRATOR: My name is Al. My name is John. My name is Lou. My name is you. I'm C.D., chemically dependent, hooked on drugs. Hooked, and I *mean* hooked. Its like being sucked into a pit of quicksand. My drug is alcohol. It could have been any number of other things. I chose that one, chose it because it was there, it was available, it was acceptable. And I got hooked!

I'm not a bad person, believe me, I mean that. Not bad just hooked. I got sick and those around me got sick too. The ones I loved the most got infected the worst. My life was pieces--non-attached, non-related, non-sense! Pieces orbiting around the only thing I really loved--my chemical.

Judge me, if you will, if you dare. I'm not worried. I don't care. The judgment brings guilt and shame. Who is wrong and who's to blame? What difference? NONE! For I was sick and if not treated, not just sick but dead!

My life began with an end--an end to the use of that chemical. No more! Just no more! Then repatterning, re-programming my life around other things, other people. I found people who cared, people who loved and with that discovery, my life began. It didn't just start, dear friend, it began!!

(MUSIC: UP FIVE SECONDS, FADE UNDER).

(SLIDES CONTINUE)

NARRATOR: Our family came apart. There we were, your average, typical middle-class American family. A home,

VIDEO	AUDIO

(SLIDES CONTINUE)

a station wagon and 2.3 kids. Everything you could want, or so we thought.
Want! Now how ridiculous is that?
What did we want? Another car, a
boat, a summer home. More things to
surround our already inflated "thing
collection?" Then it came apart, all
of it. More and more we spent time
at each other's throats, rather than
at each others' sides. We competed--
mother against father, father against
son, mother against daughter, husband
against wife. It's amazing how many
ways you can play the game. Anything
you can do I can do better. Not
better, but *bitter.* Because that's
what we became. Oh Lord, how slowly
it all happens, like a cancer which
starts out with one tiny deformed cell
and soon consumes everything around it.
We came apart. But in that process
we began asking ourselves and each
other whether we had really ever been
together. Under one roof, eating from
one table. Under one roof when we
weren't able--to be somewhere else!
But together? Never! And what does
"together" mean anyway? We began
asking, and in that asking began
finding--the answers, and each other.
We are still broken, fractured, fragmented. But we're beginning to heal.
Life has begun. Not same life, re-
treated, same meal reheated. But
new life, begun and beginning.

(MUSIC: UP FIVE SECONDS, FADE UNDER).

NARRATOR: That's my home. It *was*
anyway. For eight years I lived in
that world, or would "survived" be a
better description? I guess it would.
That was my owrld, 5 by 9, sink, toilet
and two beds. I got there because I

117

VIDEO	AUDIO

(SLIDES CONTINUE)

did something which endangered society.
What specifically doesn't matter. It
could have been holding up a liquor
store, stealing a car. Rape, robbery!
Now don't feel sorry for me. In the
name of God, don't feel sorry, because
it was pity, *my* pity, for *me*, which
put me in that world in the first
place. I did something I should not
have done. I can't blame my family,
my church, my society or my school.
I carried the gun, I performed the
deed and I paid the *price*. For eight
years my life was regimented and regu-
lated--one day just like the last,
and every day like the next. Long
periods of boredom punctuated by brief
periods of nothing to do. But I am
out now. I walk among you. I sit
alongisde you. I ride the same bus
and walk the same street. I am back
and I want to begin. If I only start
I know where I'll be. I know I'll see
those walls again. I need to begin.
I think I can do it, in fact, I *know*
I can do it. Help me! Help me begin.

(MUSIC: UP FIVE SECONDS, FADE UNDER).

NARRATOR: I am old. The better part
of my life for me has been--
The baby's cry the childhood grin
Are for me
But memory.
My years have been three score or four.
I know there can't be too much more.
Yet age is life and not a sin.
I too have cravings to begin.
Believe that, friend, with all your
heart,
I do not want a life apart.
I am human just like you,
A tad bit older that is true.
And each day can be new for me

VIDEO	AUDIO

<table>
<tr>
<td>(SLIDES CONTINUE)</td>
<td>A hand to touch, a color see.
I did not die at 65,
But I'm still very much alive.
I am old, I am not dead,
And it is not the grave I dread,
But lonliness and deep depression,
Thoughts, ideas sans expression.
My beginnings are quite small.
Yet for me they are my all.
Sunrise, snowflake, change of season
Each give me a special reason
To live that day,
Those steps to trod.
And at day's end, give thanks to God.</td>
</tr>
<tr>
<td>CUT TO STUDIO
   MCU "PREACHER"</td>
<td><u>PREACHER</u>: (VERY PIOUSLY). Amen!
Amen!</td>
</tr>
<tr>
<td>DISSOLVE TO SLIDES</td>
<td><u>NARRATOR</u>: No, no, no. Not Amen.
That's an end. We're talking about A
beginning, yours and mine. Possi-
bilities to realize, potential to
actualize. Our beginnings. Every
day, every hour, every minute. Nothing
wasted! Nothing ever wasted, not since
the beginning. Unless, of course,
we choose to waste them. And that
would be a shame, a sin in fact. Both
for you and for me. Don't waste,
please don't waste those precious,
beautiful beginnings.</td>
</tr>
<tr>
<td>CUT TO MS
   (MAN IN BATHROBE)</td>
<td><u>MAN</u>: Well, here's to beginnings!</td>
</tr>
<tr>
<td>CUT TO MS
   "PREACHER"</td>
<td><u>PREACHER</u>: Yes, here's to beginnings!</td>
</tr>
<tr>
<td>CUT TO SLIDES</td>
<td><u>NARRATOR</u>: Well, then, are you ready.
It's all out there. It really is.
So, quite literally, my friend, for
God's sake, BEGIN.</td>
</tr>
</table>

VIDEO   AUDIO

                                         (MUSIC: "CREATION" THEME, UP AND HOLD
                                         TO PROGRAM END).

FADE TO BLACK   (MUSIC: FADE OUT).

(PRODUCTION NOTE: While the script above is single-spaced, the normal
video script is double-spaced (sometime triple) for ease of reading
and to relate video instructions clearly to a particular audio section
of the program. A slide, quite obviously, is chosen to exactly fit
a line of narration).

Children's Video

     Cable presentations specially--produced for children we see as
a particular need and opportunity. Many of us despair over the
quality and content of video fare currently offered our children--
and the values underlying much of the television to which our young-
sters are exposed. Even before they can read, children are taught
the "virtues" of consumerism--the need to need, to want more and more,
and even more!

     But, the coming of cable presents congregations and local church
clusters an opportunity to do some creative video events aimed at
the child and to counter some of the crass and violent offerings of
the tube. Local thespians are often capable of stunning children's
drama. Informational pieces on local personalities, historical sites
or current issues can be produced, using the model of CBS' "In the
News" series. In Miles City, Montana, the cable system produces a
children's show using hand-made puppets. And, while the scenery and
technology are rudimentary, the local viewership is high. Programs
using children themselves, reading original stories or showing
original art work would provide a refreshing alternative to the
mindless cartoon collage that at times seems the only youth fare on
television. The inventiveness of the child is illustrated by this
piece, one which could well be a brief vignette on a children's
video magazine.

                        Monsters and Ice Cream[9]

Joanne: I've learned so much growing up with you, as your
                 mother! One of my most favorite and usable memories

                                120

is the story we created about the monster and ice
cream.

I was so sleepy that very early morning when you came
running into my bedroom, crying

Christine: "Mommy! Mommy! I'm scared."
Then I said, "There's a monster in my dream, and it
won't go away."

Joanne: I barely opened my eyes and said. "Ask him what kind
of ice cream he likes. Since he won't go away, invite him in for ice-cream!" And I turned over and
went back to sleep.

The next morning you came back into my room, and do
you remember what you said?

Christine: "Mommy, Mommy. He likes chocolate!"

Joanne: "What?" I said.

Christine: "The monster, remember? The monster in my dream!
He likes chocolate, and he's not so bad after all."

Joanne: I wonder if that isn't the way with so many scary
things, whether it's people, or a new job we have
to do, or a new tool,

Christine: Or spiders, or monsters.

Joanne: We can get to know *someone*, or *something*, by giving
it a chance, a *choice* to be what or who it is. Instead of pushing it away we can invite it into our
lives. We may have a new friend, and we might even
change our minds about being scared.

The next time we meet something that seems scary,
let's remember

Both: The monster who liked chocolate.

(The video aspect of the vignette would, obviously, include various
shots of the mother and child, perhaps even some monster drawings,
made by the child).

121

Another source of the video presentation for the local congregation or community group would be features on film or videotape available from national or regional organizations. Churches will want to see what might be available and cleared for cable use from their denominational offices or the National Council of Church Communication Commission. Other sources include Church World Service, CROP, Religion in American Life and the American Bible Society. There may also be industrial films which are suitable or short features from local libraries for which permission to use on the cable might be secured.

One apparent problem with the video presentation used in group settings is the limited size of even the largest television screen. 21 inches of stimulation is simply not enough if the crowd is large. Once again, however, there is promise in technology-on-the-horizon. Video projectors, providing a three-by-five foot picture (or larger) and costing from $2500 to $3000 are used in many applications. Dramatically less costly will be the do-it-yourself kits soon to be on the market which will convert any television set to large-screen video for $400.

Models for Cable Presence

Video *presentations* on cable--those produced by the local congregation and those done by other public groups--offer real hope to our communities as they participate in the re-invention of television. But, in some ways, the truly exciting opportunities latent in linkage systems lies in what we have called *video presence*. We refer to cable events employing the two-way feature of the medium which can put the parties at both ends of the cable electronically "there" in each other's location. This kind of electronic interconnection is the true genius of cable as we perceive it. And some types of models of presence can be fashioned, though admittedly there is less real experience on which to base our suggested scenarios. We foresee essentially five types of event in which cable-TV could enable video presence: discussion, counseling, visitation, meetings and what we shall call community "happenings."

Video Discussion

Dialogue about public issues, about political decisions facing the community (in city council, school board, park and human rights commissions, for example), could serve to create a more informed and thoughtful electorate, we believe. This would be particularly true of discussions which move beyond the presenting group of community representatives to include the reactions and questions of the viewing public. Typically, such community events have drawn too few interested citizens--even when the issues are truly important for a municipality. People simply appear at the voting polls to vote the venture down, especially if it represents an increase in the local taxes. But cable makes possible a more effective democratic process, through a far more participatory community discussion. It should certainly give the citizenry an opportunity to see how public officials function and a better judgment when they cast their ballots.

Church discussion opportunities clearly abound. Examination of ethical questions, Bible study which includes group responses, and exploration of decisions facing the congregation--a possible building or mission project--could be the subjects for which groups in the congregation "gather" by means of cable. A model for discussion of larger world and national issues by church leaders comes from a radio series which was sponsored for several years by the Minnesota Council of Churches. We find it suitable for most cable situations, as well.

The advantages of this format are that it can be inexpensively done by local talent and that the material for discussion is in plentiful supply. Further, until we really learn some of the visual capabilities of cable-TV and how we can best use them, we shall probably rely on discussion formats as staple items for our cable-casting. Talk can be vibrant and attractive, however, as the millions of viewers who watch Carson, Cavett and Griffin will attest. Non-professional on TV can be deadly, of course, and discussion shows often are nothing but dreary. But, on the other hand, if the talk is about matters of real interest and has pace, viewers will watch--especially if the issues take on a local flavor and assert some local points of view. Again, if the discussion-- through two-way cable--allows viewer presence and input, we would expect an even more engaging event.

Let us describe the radio series in which we see a possible cable model:

> We called our program REACTION.[10] The plan for the half hour was to gather a small panel of three or four church people--men, women, lay and clergy--who gave brief reactions to events prominent in that week's news. The moderator brought a dozen or so topical items and gathered the panelists half an

hour before air-time. Together we selected
matters that seemed to promise the best discussion
and put them in a priority sequence, deciding, who
would speak first on each issue. Responses were
encouraged to be brief and pointed, lasting no
longer than thirty seconds or so. If panelists
had more than one thing to say on a single news
story, we found it more interesting if they saved
some of their response for a second 'reaction'.
This kind of discipline insured movement and
excitement often lacking in usual talk formats.

To introduce each news item, the station's news
department provided a voice actuality. Such actualities, voice recordings of persons in the news or
of on-the-scene reporters, are fed to many stations
daily from a variety of sources, including networks,
news services, regional "stringers", and the like.
Stations in your community might be willing to provide these actualities to you once a week. Actualities
should be very brief, just long enough to recall the
event for the listener, allowing the panelists to
supply the context in their remarks. If you cannot
locate actuality material, the quotation or report
could be read by the program moderator or some other
person.

A decision needs to be made by the moderator about
when to end comments on a given issue and turn to
another item chosen from the week's news (the
moderator is not really a participant in the discussion). Using six to ten items each program
provided discussion on single issues of five minutes
or less. At the open and close of the program, panelists were identified with their church affiliation,
and while "God talk" was avoided, discussants responded quite personally to local and national matters from
their own faith perspectives. The more ecumenical the
representation on the panel, the livelier and more
profitable the show tended to be.

Our suggested program aims were:

"1. To raise issues and give reactions thought-
fully, provocatively, tersely, as church men and women.
    2. To indicate that faith does speak to life.

      3. To suggest an image of the church/synagogue as a relevant institution. We do not aim to suggest easy solutions to perplexing problems; we aim, rather, to show our concern for them."

REACTION, as we did it on Twin Cities radio, served a "climate creating" function in behalf of the religious community. No single perspective nor advocacy of any ideology was sought. Through the public service time the station provided and their heavy promotion of the program it seems that others sensed that excitement as well.

Actually, any community group could adapt *REACTION* for cable use. A cable version of *REACTION* could use news photos clipped from newspapers and magazines and mounted on cardboard to provide visual enhancement to the items selected for discussion. Short clips from video recordings of TV news or footage obtained directly from a local TV newsroom is especially desirable.

Adaptations could take on a topical emphasis. For example, a program might gather the candidates for the school board elections and ask their succinct reactions to local education issues. Other local community issues, such as land use, political controversy and governmental issues offer opportunities for cable discussion.

The *REACTION* radio format looked like this:

PROGRAM OPEN

ANNCR: THIS IS REACTION. (Dramatic voice.)

MUSIC: Up strong for 5 seconds, fade and hold under Anncr.

ANNCR: Reaction, a frank discussion of vital issues, is a public affairs presentation of this station in co-operation with *the Minnesota Church Committee on Radio and Television (*here insert name of your local group or community agency).

MUSIC: Up full for two seconds, fade under Moderator and out.

MODERATOR: OUR REACTORS THIS EVENING: (Here each panelist
                introduces himself by name and church affiliation).
                And this is your moderator (give name).

                I ask for your frank reactions to the matters before
                us. While you speak as churchmen, we recognize you
                do not speak officially for any church body, nor are
                the opinions expressed necessarily those of the
                management or staff of this station. Let's begin
                with this quotation out of the recent news:

(In the body of the program, the Moderator continues to ask for
responses to individual news items, occasionally mentioning the
program name in transitions from item to item.)

PROGRAM CLOSE

MODERATOR: Thank you for these reactions. Our reactors this
                evening were: (Here each panelist indicates name
                and church affiliation again.) And this is your
                Moderator (give name).

ANNCR: This has been reaction.

MUSIC: Up strong for 5 seconds, fade and hold under Anncr.

ANNCR: Reaction, a program of response and comment on matters
                important to everyone.

MUSIC: Up strong for 2 seconds, fade and hold under Anncr.

ANNCR: Reaction, a public affairs presentation of this
                station in cooperation with: (Here insert name of
                local church group or community agency).

MUSIC: Up full to program close.

Video Meetings

In April, 1979 linkage technology, in this case relying solely
on satellite broadcasts, enabled a *video meeting* of unusual proportions. The *World Symposium on Humanity* was held in three locations,
Los Angeles, Toronto and London, with participatns involved inter-
nationally in common stimulation and interacting by means of video.

Sponsors of the symposium wrote of their project in this way:

"Human society is passing through a traumatic renewal.
We are past the time for mutual recrimination, and it
is essential that we face certain well-evidenced dangers.
We need a new realism. Even now, as the psychological
and physical fabric of our civilization begins to unravel,
new resources of awareness, communication, and technology
emerge. The World Symposium on Humanity will celebrate
that emergence, with activities designed to promote
healthier lifestyles, and with contributions from people
in all stations in life committed to constructive,
humanitarian action. Local, regional and global connections will be made among diverse groups of people, in
an attempt to affirm the existence of, and give direction
to, a powerful new consensus of evolutionary proportions."[11]

Less gradiose but perhaps often as vital for local communities
would be the revival of the old American tradition of the town
meeting, this time electronically convened. It would give citizens
an opportunity to be heard on the issues and could include some
people--the handicapped and aged--who have been virtually excluded
from public meetings in the past. Cable gatherings would have the
same disadvantages as the old town meetings--boring speakers,
long-winded speeches--but the technology would also control by the
switch time limits formerly difficult to enforce.

Recently a suburban church in mid-America experienced some
staff changes. The resignation of two clergypersons provided the
congregation an opportunity to step back, assess its staffing needs
and come at the task of replacing the persons with a better idea
of what they wanted for their church. The study which was done
resulted in recommendations which had major implications for the
future, both in terms of staff configuration and budget. It was
essential (if the proposed plan was to work) that the congregation
be fully informed, not only about the recommendations but the background and basis upon which the recommendations were made. The
only available way was through a special congregational meeting
on a Sunday morning following the worship service. While the
attendance was good, those present represented only about 40% of
the total membership.

With a wired community, on the other hand, it could be
announced for several Sundays that the church was having a special
cable program on the matter. People would be encouraged to watch,

127

and using the two way capacity of the system they could register reactions to the proposals from their homes. The decision might be made to repeat the cablecast on another night to make it possible for more members to receive the information and register their opinions. What parish minister would dare suggest holding two congregational meetings in one week? There is, of course, no way of telling how many people would take part in such a church "meeting." The video pencil does not *guarantee* anyone will read what it writes. It does however make information more accessible to more people and provide, through its upstream capability, a sense of participation which no current communication medium can match. We see this kind of participation enabling meetings of other kinds and involving groups of many sizes--all the way from denominational committees and task forces (now involving costly travel and time-sacrifice as people gather from around the nation) to the Christian education committee of a local congregation. A choir practice would seem to require the usual physical presence of its members, but who would finally limit the possibilities of electronic linkage?

As the flow of information is facilitated and persons can participate in decision-making in new ways, the very strong possibility exists of strengthened communities. Groups of persons may well be drawn closer together by a genuine sense of ownership in what the organization is doing and dreaming. This free flow of information *in both directions* is particularly critical to the life and health of organizations such as the church. Many congregations are built on the consensual model, yet often function on less than true consensus. The church, in particular, embraces a polity which would affirm all and exclude none, yet practically its functions are often *exclusive*. But, cable presence gives us another possibility.

Konzelman puts it this way:

"Since with two-way communication potential, small groups in houses or other remote locations can be tied together for intercommunication with other small groups, a sense of identity with the larger community could be maintained and even heightened while maximizing the possibility of indepth communication between a few people at each location. Such a facility would provide every opportunity for active participation and interaction now available through the usual type of meeting or learning situation, plus new and yet unplumbed forms

of activity and relationship. In such a situation,
the "wired church" could be a reality and the
coaxial cable would become the link to deeper and
more meaningful church community."[12]

Video Visitation

A less obvious possibility for cable presence, one about
which we anticipate early clergy ambivalence, is the *video visitation*. We suggest here that the linkage technology which will
allow a person to play chess with another several hundred miles
away by audio-video interconnection could certainly be used in
other kinds of person-to-person contacts, including those which
would enable the ministries of the local church. In other words,
we foresee the day when some of the minister's calls in hospitals,
nursing and retirement facilities, and the homes of parishioners
will be made by means of the cable.

A minister acquaintance of ours regularly makes a 64-mile
trip in order to visit a particular shut-in. One and a half hours
and three gallons of gas are expended, in the midst of a busy
pastoral schedule. In actual fact, more time might well be spent
with that same parishioner if the minister could conduct the visit
by video, yet with precious resources in time, money and energy
conserved. Another minister attests to the fact that the telephone currently enables significant person-to-person pastoral
visitation. The medium makes possible both private and intimate
sharing, bringing minister and lay-person into each other's
presence in an important way. When the technology adds a high
fidelity picture of the other to our audio communication, the
reality and impact can only be heightened. Such visitation
will not replace the literal visit, of course, nor should it. But,
there is much to interest us in video visitation in a time of
energy shortage and increasing encroachment on the time and
strength of the clergy. Further, the cable visit may be a ministry
taken up by more lay people when they can accomplish it with less
time and dislocation.

Video Counseling

A fourth genre of video presence is *counseling* by means of the
cable. A special need in this instance is the maintenance of abso-
lute privacy and our interest in the opportunity is based on the
assumption that privacy can be assured. With such security, we can

envision counseling--of couples planning marriage, of the chemically dependent, of the despondent and the grieving--all going on
by two-way cable facilities. The minister may not be able to reach
out to take the hand of the one who is grief-stricken, but her
presence, in some ways, may be profoundly *real*.

Video "Happenings"

There would be yet other kinds of community events which would
merit use of a cable system to create *presence,* events which--for
convenience--we would label "happenings." We think of neighborhood
festivals and carnivals, of art shows and parades, picnics and craft
displays. And, again, we see new opportunities for people--especially
those not easily able to take part physically--to "be there" via the
electronic magic of cable. We have learned very quickly how to stage
football games and political conventions for television. Now, perhaps, we shall learn to manage community events in order to convenience those participating from their homes or places of business.

Our models of cable presence, each made possible by two-way
linkage systems, illustrate our thesis that ministries traditionally
taken up by the church can be enabled in important new ways. If we
begin to understand more fully this promising feature, then cable
communication may lose its apparent threat to clergy and congregations and may excite more active church support in its development.

Clearly, the issue must be faced of financing a church or community group's involvement in cable use. There will be costs,
potentially substantial costs, in purchasing equipment, in producing
tapes, perhaps in leasing time on a special channel. On the other
hand, as we have envisioned, linkage systems will function as
electronic transportation systems, "bringing church conventions,
councils and other conferences to the constituency, rather than
having to bring the constituency to the conference or convention."[13]
The savings in travel, hotel and restaurant expenses would amount
to millions. In the congregation, there would be money saved on
church building and maintenance expenses, freeing up funds for its
ministries in cable presentation and presence. In other words,
cable communication may not so much mean *new* expenditures, as the
diversion of items on the present budget.

Strategies for Cable Ministry

Many of those who have written about the technology and its

implications for church and community have suggested plans and procedures to be initiated now. Earlier, we added our voice to those urging immediate attention to the issues brought us by the medium of cable, before its full impact is felt--while linkage systems are malleable and under *our* control.

One to propose a strategy--an interim plan before the cable comes down the neighborhood street--is Nelson Price.[14]

"1. If you now have cable, visit the manager. Find out what he is doing in local programming. Discuss how you might help provide programs.

2. Do not wait for cable. Begin to use video in your education program, with your choir, in your community.

For example, First Presbyterian Church in Atlanta, Georgia, has used its video tape recorder with its high school class in producing dramatic episodes. They have used the tape for leadership training and skills training of teachers. Now six Presbyterian churches in Atlanta are planning to pool equipment, specialized talent from each congregation, and tapes....

3. Move out into the community. Share your equipment and expertise. Become acquainted with community video workshops. The Alternate Media Center of New York University has helped establish continuing workshops in Cape May, New Jersey; Reading, Pennsylvania; Columbus, Indiana, and DeKalb, Illinois.

4. Create a feedback system. Your production group will lose interest if it does not secure reactions to what it is doing. Your first response can be from the persons you've recorded. Let them see it. Let them know when it will be on cable. They will tell their friends. But, another important mind-set shift, you will not inherit a large audience from the preceding program as on broadcast television. The audience will not be there unless you get them there."

Musing particularly about the role of the local church in cable-television, Robert Konzelman suggests these steps in how to begin:

"1. Learn everything possible about cable technology;

2. Become informed about the status of cable in the local community: what is presently and po- tentially available;

3. Participate in local community efforts to develop cable and to insure the community's best interests;

4. Begin to think and plan for possible utili- zation of cable, even on a limited basis, making use of present existing opportunities in the light of the goals and purposes of the parish;

5. Begin to develop familiarity in the parish with the visual media, especially video tape equipment, etc., using video programs and equipment for in-house program development by children and youth and adults;

6. Experiment with various forms and structures for utilization of existing and forthcoming cable programming."[15]

Approaching the same issue--of designing strategies to prepare for the cable age--this time from the point of view of public school educators--Welby Smith puts forth these suggestions, abbreviated here for succinctness.

"If at all possible, the school system should be involved in the franchise award process from the beginning. Remember, letting the franchise to a profit-making entrepreneur isn't the only course. Local non-profit groups, institutions, or the county or city itself could put in and run the system....

Don't be greedy when it comes to channels. Justify your requests for channels with hard-nosed program planning....

Don't let the Cable Operators, Broadcast Industry or anyone else close out half-inch and one-inch VTR origination. [Today, we would want to add 3/4 inch cassettes]. These constitute the only means that

132

the public, and most teachers, will ever have of
developing and producing programs....

Free up access to program origination equipment
within your own system. Teach your teachers to
use the equipment, not to fear it....

Form a partnership with the public. Provide
training, access, advice. Don't shut out anybody,
or allow anybody to be shut out...by appeals to your
philosophy, vanity or expertise.

Once you get your channel(s), make sure you also
have control of the production process. Input,
even content control, is not enough. You need
nothing less than the capability to develop and
produce your entire program, down to a final edited
tape....

Explore the concept of a "Media Ombudsman" [sic],
someone independent of the CATV operator who can
furnish necessary technical and production assistance for systems that don't have in-house resources....

Fight for maximum freedom of expression in programming...."[16]

Perhaps, as one city official once remarked, "If the public
can be aroused, there is a chance we can prevent undue skullduggery
from transpiring in the closets." The strategies suggested here,
we feel, will help in such arousal. We see them as viable plans to
awaken interest, to educate, to equip and to mobilize the church
and community. Where franchises are still to be written prior
public consciousness-raising will certainly result in the framing
of documents better insuring public service from the cable. In
communities where systems have already been franchised and are
functioning, public awareness can bring about ampler exploitation
of the cable for the good of the community.

Conclusion

Our potpourri of models and strategies is before you. The
collection is fragmentary and unfinished. One reason for that is
the authors' creative finitude! For this moment we have shared

our ideas. But, another rationale for limiting the models is that responsible use of cable presentations and presence can finally be designed only in its community context. The genius of cable-TV, as we have insisted, is in its localism. Its best use is shown in software tailor-made for its immediate locale by those who live there.

We have not really tried to solve the problems of those who want to create new human networks via cable-communication. On the other hand, we *have* tried to rob video of some of its mystery and mystique. As one has said, recording a problem on videotape can reduce it to the size of a TV screen! That can make the issue more manageable and encourage people to work on solutions. We would like to convince folk they can stand up to TV--before the camera as talent, or behind it as producers. In a word, we have been working to sharpen the video pencil--ours and yours.

On the other hand, we have had to leave untouched many issues of real importance. And there is need seriously to address such questions as the impact on linkage systems of energy shortfall, of economic factors and how they will influence the development of cable and its service to the public, of community access to cable channels, of franchising considerations, of privacy invasion, of free speech and censorship, of preparing the populace for spectacular choices, of insuring fairness on the cable for minorities and women. The list is long, and it itemizes crucial matters. The scope of this work and its basic purpose, however, have deterred us from lengthier discussion.

Our point, in truth, has been to raise the consciousness of a people we know well--the People of God, gathered across the land in the parish churches, and their friends in the neighborhood. And, as we have affirmed, it is precisely this institutional expression of the Judeo-Christian community which possesses the most immediate opportunity to influence and use the emerging technology. We have insisted that a cable system can function especially to facilitate and enable those ministries of the congregation which are traditional and long-perceived--preaching, teaching, nurturing and caring.

But, the lure of linkage systems summons us also to be experimental, to be creative and to be far-thinking. For, the church is surely called to dream the dreams of more fulfilling life and to see the visions of a reconciled human community. It is called to *speak* --a gracious and healing word, but often in a new guise. And it is elected to *serve* --a public chafing from injustice and

unused to electronic caring. When the congregation's speaking
and serving is done by means of the cable, we shall want to
employ our best. We shall need wit and inventiveness and perse-
verence; we shall have to be hard-working and hard-nosed. Cable
communication *will bring change*, in truth. It will not be another
television "repeat" for either church or community. Clergy and
laity will want some re-telling. And we would call the church and
the churches to take up the burden of the possibilities.

How, then, shall we speak of the new delivery system? Is it
a monster, invading our privacy, bringing a plague of new ills to
a world already in fragile health? Or is it the TV of abundance,
a marvelous electronic highway, a communications cornucopia? Will
the cable bring us cultural mediocrity or societal enrichment?
Our choice is to speak of cable-TV as the *video pencil*. It will
write good scripts and bad. But it is given us in our time to help
each other as we take that pencil in our hand.

NOTES

[1] Chuck Anderson, *Video Power: Grass Roots Television* (New York: Praeger Pubs., 1975), p. 11.

[2] *Ibid.*, pp. 11-13.

[3] "The Church and Its Potential in Cable Television," an unpublished article distributed by the Division of Parish Education, the American Lutheran Church, Minneapolis, Minnesota, May 10, 1972, p. 5.

[4] *Ibid.*

[5] The script was written by Louis G. Wargo and was telecast on KTCA-TV, St. Paul, Minnesota, June 12, 1973.

[6] Guy Stone is on the clergy staff of Parkway United Church of Christ, Minneapolis, Minnesota, and presented "Real People" at the church during Lent, 1980.

[7] This material was written by Louis G. Wargo and presented at the 1978 Annual Meeting of the Eastern Association, the Minnesota Conference, UCC.

[8]The video script was written by Louis G. Wargo and telecast on WTCN-TV, Minneapolis as its *Sunday Morning Worship* program, on July 18, 1976.

[9]The original script was conceived and written by Christine and Joanne Perrin of Brooklyn Center, Minnesota. Joanne is a student at United Theological Seminary of the Twin Cities and Christine is her six-year-old daughter.

[10]*Reaction* was produced by Gene Jaberg for broadcast on WTCN, Minneapolis and for many years was sponsored by the Minnesota Council of Churches. We have used here materials from Jaberg's "Citizens 'React' to Current Issues," in *The Originator* (Sept./ Oct., 1973), p. 6.

[11]Quoted from a brochure promoting the *Symposium* and distributed by the *Humanity Foundation* of Vancouver, B.C.

[12]*Op.cit.*, p. 4.

[13]*Ibid.*, p. 6.

[14]"What Cable Can Do For You," in A.D. (May, 1973), pp. 39, 40.

[15]"The Church and Its Potential in Cable Television," in *Learning With Adults in the Parish* (July-Aug., 1973), p. 32.

[16]"You, Me and Cable-TV," in *Media and Methods* (Feb., 1973), pp. 18, 19.

136

A SELECT GLOSSARY OF CABLE TERMS

access channels - Free non-commercial channels open for public use on a non-discriminatory basis. No longer guaranteed by FCC provision, but--from our point of view--a legitimate claim to be made by church and community groups of the system operator.

broadband - Has reference to the broad-ranging kinds of information services carried by linkage systems. Cable television is sometimes called broadband communication.

CATV - Stands for Community Antenna Television System, a means by which television signals are received and re-distributed to multiple receivers. Another way of referring to cable-TV.

cablecasting - Local programming by a cable system, distributed by cable rather than being broadcast over-the-air.

chroma key - A special video effect, in which one figure can be matted over another background. The effect is achieved by using the color camera's blue camera signal.

closed circuit - Transmission of video signals, normally by a cable, to reception points equipped with TV monitors. *Open* circuit would refer to broadcast TV in which signals are transmitted indiscriminately.

coaxial cable - The common wires used to distribute cable signals. It is constructed of an outer conductor or shield which surrounds a center conductor which is held in place by an insulating material.

conceptual video - TV programs in which video electronics are employed to create artistic events.

distribution system - All the necessary electronic equipment required to convey signals from the head-end of the cable system to subscribers' sets.

drop - The section of cable which taps off the feeder lines and delivers the cable signal into the home or other place of reception.

facsimile - Information in print which is translated into electronic signals and carried by CATV to the subscriber. There, on special equipment, the signal is converted into a print-out.

franchise - The agreement between a cable operator and local government which states the responsibilities and rights of each party for the construction and operation of a cable system in a given community.

hardware - The electronic equipment by which the "software" is transmitted. This, of course, would include studio cameras, lights, special effects generators, film chains and the like.

head-end - The location of the main reception equipment for a cable system. There one will find the needed components to translate channel frequencies, adjust sound and picture, compensate for unequal cable losses and switch to different channels the various sources of programming.

leased channels - Refers to channels available for a fee to interested groups or individuals, who--in turn--may cablecast whatever materials they wish, so long as it is in accord with FCC regulations.

linkage system - Another way of describing the function of cable, one recognizing that the literal means of signal delivery may involve copper cable, fiber optics, laser beams, satellites or technologies yet-to-be.

message wheel - A device which rotates small cards displaying messages or advertising in front of a TV camera. Message wheels commonly turn throughout the day, providing the automated programming for a special channel.

narrowcasting - In contrast with "broadcasting," an alternative term for cablecasting which stresses the smaller, often more specific audiences for which programs are designed.

pay cable - The provision of special programs via le sed channels and available to the subscriber for the payment of an extra fee.

portapak - A complete but portable video rig, including camera, video tape recorder, battery pack, etc. The equipment is miniaturized and often is hand-held.

software - Refers to the program content, the material shown on television broadcast or cablecast. Includes what is recorded on films and videotape, or what plays out live from other sources.

street video - Programming for cable-TV dealing with issues and events important to a community, originating from the community--off the street, as it were.

subscriber - The person whose television receiver is connected to a cable communication system. The cable programming is available for a monthly fee, and one "subscribes" to the service.

trunk line - This is the main cable of a system, feeding the signals from the head-end to the remainder of the system's lines.

two-way system - A cable system which can both transmit information from head-end to subscribers and receive information from any point in the system.

upstream capability - Refers to the two-way feature, above. The capacity of cable systems to carry signals, as it were, "upstream" from the receiving set to the originating source, the head-end or ultimately to other receivers.[1]

NOTES

[1] We have made reference to some glossary terms in Don Schiller, Bill E. Brock and Fred Rigby, *CATV Program Origination and Production* (Blue Ridge Summit, PA, TAB Books, 1979), pp. 236-249. For a more complete glossary we recommend this extensive listing. We find the book as a whole useful on the subject of cable production and management.

INDEX

Guinness, Os - 62
Gutenburg - 3, 4, 5, 6, 42, 43

Haight, Timothy R. - 32
Hartshorne, Charles - 47, 48, 56
Haselden, Kyle - 54, 71, 72, 73
Henry, Nelson B. - 56
Herzog, Frederick - 43

IEEE Transactions on Communications - 31
Institute for Visual Learning, Inc. - 86
Irenaeus - 55

Janare, Richard Paul - 85
Jennings, Ralph - 31, 78
Johnson, C. Edward - 88
Johnson, Nicolas - 28
Jones, James C. - 31

Kabla, Kas - 7, 13, 24, 33, 99
Kaplan, Abraham - 57
Kaplan, Donald M. - 64
Kearney, Lawrence - 57
Kegley, Charles W. - 56
Kirk, Kenneth E. - 54
Konzelman, Robert - 32, 61, 85, 99, 128-129, 131-132
Kraemer, Hendrik - 38, 54

Lieberman, Shirley and Dean - 86
Lippman, Walter - 44
Logan, Ben - 70
Lutheran Standard, The - 32

McKay, Patricia - 87
McLuhan, Marshall - 5, 11, 23, 37, 51, 52, 57, 61, 63
    -global village - 11
Martin, James - 32
Matson, Floyd W. - 56, 57
Mead, Margaret - 54, 80-81
Meland, Bernard - 48, 56
Mill, John Stuart - 74
Milton, John - 74
Minnesota Council of Churches - 123
Moody, Kate - 86

GEOGRAPHICAL EXPERIMENTS IN CABLE TELEVISION

A) Early Experiments
      Great Britain - 17
        Gloucester - 17
      Pennsylvania
        Lansford - 17

B) Church Experiments
      New York
      Connecticut
        Waterbury - 77
      Indiana- 83
      Tennessee
        Knoxville - 83
      Georgia
        Augusta - 83
        Atlanta - 131
      Wisconsin
        Madison - 83-84
        Bloomer - 84
      California
        Long Beach - 84
        Castro Valley - 84
      New Jersey
        North Bergon - 84
      Texas
        El Paso - 84
      Illinois
        Peoria - 84
      Oklahoma
        Tulsa - 84
      Missouri
        St. Louis - 84
      Pennsylvania
        Philadelphia - 83
      Virginia
        Richmond - 83

C) Two-way Experiments
      Ohio
        Columbus - 6, 29
      Virginia
        Reston - 22, 29
      Japan
        Toma New Town - 29

D) Management and Regulation
  Minnesota - 19
    St. Paul - 77
  Massachusetts - 19
  Mississippi
    Jackson - 79

BIOGRAPHICAL NOTES

Gene Jaberg is Professor of Communication at United Theological Seminary of the Twin Cities, New Brighton, Minnesota. Born in Linton, Indiana, he studied at Princeton Theological Seminary and at Westminster Theological College and Cambridge University in England, and has degrees from Lakeland College and Mission House Theological Seminary. At the University of Wisconsin (Madison), he was the H. V. Kaltenborn Radio Scholar and was awarded the M.A. and Ph.D. in Speech Communication. Mr. Jaberg has been a staff announcer and writer at several Midwestern radio and television stations, served as an army correspondent, was minister of Pilgrim Congregational Church in Madison, and since 1958 has been a theological educator. He was managing director of Interfaith Players for seven years, has produced a number of films and television programs and has written articles and book reviews for a variety of publications. Mr. Jaberg has been active in the Speech Communication Association and the Religious Speech Communication Association, and currently serves on the United Church of Christ Office of Communication Board of Directors.

Louis G. *Wargo* has been the senior minister of the Congregational Church of Excelsior, Minnesota since 1972. He has served pastorates also at Alpena, South Dakota and Grand Marais, Minnesota. Mr. Wargo has worked in radio, written and produced television programs and films, and has chaired ecumenical church communication committees. In 1973 he received a grant from the Fund for Theological Education (supported by the Rockefeller Foundation) to study cable communication and its implications for parish ministry. Born in Norwalk, Connecticut, Mr. Wargo has degrees from Franklin and Marshall College and the Divinity School of Yale University. He has done graduate work in radio and television at United Theological Seminary and the University of Minnesota. With Mr. Jaberg, he is co-producer of two 16mm motion pictures, "They Call It Cable," and "New Medium: New Messages."